H 976

Praise for *ANYWHERE*

"We ain't seen nothing yet! That's my conclusion after reading this inspiring, thought-provoking, and insightful book. Although everyone knows that the ICT industry can be driven by hype, creating booms and subsequent busts, it is difficult to 'over-hype' this book, since it describes an inevitable transformation that business must go through. The sooner you understand this, the better off you will be. Emily's inspiring and insightful book is about a journey into this 'connected' future. After reading it, I can't wait to get on the road and am certain to bring Emily with me as my guide."

—Michiel Boreel, CTO, Sogeti

"*ANYWHERE* paints a compelling picture of what the next transformation of wireless will look like and who it will impact. Those who want to capitalize on the new wireless world should read this book."

—Dan Hesse, CEO, Sprint/Nextel

"Emily Green's *ANYWHERE* is required reading for anyone interested in understanding how and why communications advances are fundamentally altering business. Green's interesting case studies—each a story unto itself—examine the Anywhere impact already being felt by consumers across the globe. Overall, she explains how the connectivity revolution offers unbounded opportunity to thrive in the nascent Anywhere future."

—Reed Hundt, former chairman, U.S. FCC

"The acceleration of connectivity worldwide is perhaps the most crucial new dimension driving and changing all businesses. Emily Green correctly identifies that while latency kills commerce, meeting customer expectations for performance drives growth. Anywhere is a revolution, and every manager in every industry must understand the Internet insights captured in this book."

—Paul Sagan, president and CEO, Akamai

"I've long regarded Yankee Group as the preeminent resource for insights into connectivity. What distinguishes their vision is the richness of data they collect and the rigor they apply to understanding it. In this book, Emily Nagle Green takes it up a notch; with a breezy style that makes the arcane accessible and the possible plausible. It's the kind of sensible outlook that can only come from deep knowledge and analytic rigor, both hallmarks of Yankee Group."

—Tom Sebok, president & CEO, Young & Rubicam North America

"Connectivity is fast creating a level playing field among developed and emerging markets. Those who understand how to leverage the connected world will be best positioned to impact it. *ANYWHERE* is a must read for anyone who wants to be a relevant leader in a global economy."

—Rajeev Suri, CEO, Nokia Siemens Networks

"This book highlights the unstoppability of the advance of connectivity, creating the imperative for business leaders to respond."

—Ben Verwaayen, CEO, Alcatel-Lucent

"I've participated as an entrepreneur in the evolution of mobile and broadband technology and have witnessed firsthand the huge impact they have had, and will have, in business and society. Green's book taps into all of the most important issues that will impact these areas in the near term and the future. With this book she's compiled the voices and opinions of the most influential people in the category for a great read that I highly recommend."

—Larry Weber, chairman and CEO, W2 Group, and author of
Marketing on the Social Web **and** *Sticks and Stones*

ANYWHERE

How GLOBAL CONNECTIVITY
Is Revolutionizing the Way
We Do Business

EMILY NAGLE GREEN
President and CEO of Yankee Group

New York   Chicago   San Francisco   Lisbon
London   Madrid   Mexico City   Milan   New Delhi
San Juan   Seoul   Singapore   Sydney   Toronto

1 2 3 4 5 6 7 8 9 0   WFR/WFR   0 1 0 9

ISBN: 978-0-07-163514-1
MHID: 0-07-163514-9

This publication is designed to provide accurate and authoritative information in regard to the subject matter covered. It is sold with the understanding that the publisher is not engaged in rendering legal, accounting, or other professional service. If legal advice or other expert assistance is required, the services of a competent professional person should be sought.

—From a Declaration of Principles Jointly Adopted by
a Committee of the American Bar Association and
a Committee of Publishers and Associations

McGraw-Hill books are available at special quantity discounts to use as premiums and sales promotions, or for use in corporate training programs. To contact a representative, please visit the Contact Us pages at www.mhprofessional.com.

This book is printed on acid-free paper.

Library of Congress Cataloging-in-Publication Data
Green, Emily Nagle.
  Anywhere : how global connectivity is revolutionizing the way we do business / by Emily Nagle Green. — 1st ed.
     p. cm.
  Includes bibliographical references.
  ISBN 978-0-07-163514-1 (alk. paper)
  1. Globalization—Economic aspects. 2. Telecommunication systems. 3. Internet. I. Title.
  HF1359.G724 2010
  658'.054678—dc22

                                                    2009031990

ANYWHERE

How GLOBAL CONNECTIVITY
Is Revolutionizing the Way
We Do Business

EMILY NAGLE GREEN
President and CEO of Yankee Group

New York   Chicago   San Francisco   Lisbon
London   Madrid   Mexico City   Milan   New Delhi
San Juan   Seoul   Singapore   Sydney   Toronto

The following are registered trademarks ® of the Yankee Group: the Yankee Group name, the Yankee Group logo, WiMax World, Anywhere Consumer, Anywhere Enterprise, and Anywhere Network. The following are trademarks ™ of the Yankee Group: 4G World and Mobile Internet World.

1 2 3 4 5 6 7 8 9 0   WFR/WFR   0 1 0 9

ISBN: 978-0-07-163514-1
MHID: 0-07-163514-9

This publication is designed to provide accurate and authoritative information in regard to the subject matter covered. It is sold with the understanding that the publisher is not engaged in rendering legal, accounting, or other professional service. If legal advice or other expert assistance is required, the services of a competent professional person should be sought.

—From a Declaration of Principles Jointly Adopted by
a Committee of the American Bar Association and
a Committee of Publishers and Associations

McGraw-Hill books are available at special quantity discounts to use as premiums and sales promotions, or for use in corporate training programs. To contact a representative, please visit the Contact Us pages at www.mhprofessional.com.

This book is printed on acid-free paper.

Library of Congress Cataloging-in-Publication Data
Green, Emily Nagle.
  Anywhere : how global connectivity is revolutionizing the way we do business / by Emily Nagle Green. — 1st ed.
      p. cm.
  Includes bibliographical references.
  ISBN 978-0-07-163514-1 (alk. paper)
  1. Globalization—Economic aspects. 2. Telecommunication systems. 3. Internet.
I. Title.
  HF1359.G724 2010
  658'.054678—dc22

                                                    2009031990

CONTENTS

Part III The ANYWHERE Enterprise

Part IV Profiting from ANYWHERE

The Internet Comes of Age

By Don Tapscott

As Emily Nagle Green explains in this breezy yet insightful book, there is a new medium of human communications emerging that is bringing deep changes to the way we create value in society, do business, work, live, and play. The Internet has finally come of age, and the result is a world of Anywhere.

Emily outlines a fascinating paradox that holds great promise. Location matters less than ever before: We're all on a network and can access whatever people, resources, and information we need to work, learn, or play. But location also becomes more important than ever before because we are all being freed to be where we want to be. Personally, in the last year I've been connected to the rest of the world working from my house on a remote lake in Ontario, Canada, in the Masi Mara wildlife reserve in Kenya, and on a beach in Bali.

Billions of connected individuals can now actively participate in innovation, wealth creation, and social development in ways we once only dreamed of. And when these masses of people collaborate, they can advance the arts, culture, science, education, government, and the economy in surprising but ultimately profitable ways. Companies that engage with these exploding Web-enabled communities are discovering the true dividends of collective capability and genius.

With the so-called Web 2.0, we're all participating in the rise of a global, ubiquitous platform for computation and collaboration that is reshaping nearly every aspect of human affairs.

The old Web was about Web sites, clicks, and "eyeballs." In some ways it resembled the old broadcast media more than today's new collaborative Web. As users and computing power multiply, and as easy-to-use tools proliferate, the Internet is evolving into a global, living, networked computer that anyone can program—whether building a business on Twitter, producing a video clip for YouTube, creating a community around a Flickr photo collection, or editing the astronomy entry in Wikipedia.

This new Web already links more than a billion people directly, reaching out to the physical world, connecting countless inert objects from hotel doors to cars. It is beginning to deliver dynamic new services—from free long-distance video telephony to remote brain surgery. And it covers the planet like a skin, linking a machine soldering chips onto circuit boards in Singapore with a chip warehouse in Denver, Colorado.

Twenty years from now, I'm convinced we will look back at this period of the early twenty-first century as a critical turning point in economic and social history. We will understand that we have entered a new age, based on new principles, worldviews, and business models, where the nature of the game has changed completely. And rather than as a recession, the crisis will be viewed as a reset of the global economy.

I remember my first job at Bell Northern Research in the 1970s. I worked with a team of technologists and social scientists whose task it was to understand what was then called "The Office of the Future." Our group was trying to understand how multifunction workstations connected to a vast network of networks would change the nature of knowledge work and the design of organizations.

As part of my job I traveled around the world trying to meet anyone who might know something about how emerging technologies would impact business. I was fortunate enough to collaborate with some of the pioneers of the digital age, like Stanford Research Institute's Douglas Englebart. I'll never forget him showing me his "augmented knowledge workshop," complete with hypertext, collaboration tools, and a strange device he called a "mouse." Another pioneer, Jim Bair, convinced me over a period of many months that the killer application for the future of computing was not "Information Management" as it was called back then, or for that matter computing at all—but rather communications. I also remember talking to Howard Anderson, who had founded Yankee Group, a company that analyzed where technology and the technology industry were going.

In the late 1970s at Bell, we had a pilot team of 50 managers and professionals (including our CEO) using workstations that had similar functionality to today's laptops—except that we were using dumb terminals connected to minicomputers that were in turn connected to low-speed packet-switching networks. From this experience I became convinced that collaborative computing would change the world.

But when I wrote a book on the topic in 1981, in hindsight predictably few people read it. Critics told me this idea of everyone using computers to communicate would never happen, giving the most bizarre of reasons: Managers would never learn to type. No one could have predicted that not only would managers learn to type, but they would type with their thumbs!

The ideas of Englebart, Bair, and other pioneers back then were ideas in waiting—waiting for a number of technological, social, economic, and demographic conditions to mature.

From the Internet's inception, its creators envisioned a universal substrate linking all mankind and its artifacts in a seamless, interconnected web of knowledge. This was the World Wide Web's great promise: an Alexandrian library of all past and present information as a platform for collaboration to unite communities of all stripes in any conceivable act of creative enterprise.

Today, the Net has evolved from a network of Web sites that enable organizations to simply present information to a computing platform in its own right. Elements of a computer—and elements of a computer program—can be spread out across the Internet and seamlessly combined as necessary. The Internet is becoming a giant computer that everyone can program—providing a global infrastructure for creativity, participation, sharing, and self-organization.

As Anthony D. Williams and I explained in *Wikinomics*, the hospitals and graveyards of the Web are full of sick and dying Web sites whose owners think the Internet is a platform for the presentation of content. The programmable Web eclipses the static Web every time. Think back to how Flickr beat WebShots; Wikipedia beat Britannica; bloggers beat CNN.com; epinions.com beat Consumer Reports; Upcoming beat Evite; Google Maps beat MapQuest; Facebook beat Friendster; and Craigslist beat the classified ad sites.

What was the difference? The losers launched Web sites. The winners launched vibrant communities. The losers built walled gardens. The winners built public squares. The losers innovated internally. The winners innovated with their users. The losers jealously guarded their data and software interfaces. The winners shared them with everyone.

In fact I tell my clients, don't have a Web site, it's such a dot-com idea. Rather curate a community. Build a platform for people to self-organize, collaborate, and create their own content.

I was speaking at a conference in Munich recently where my introducer said, perhaps hoping to get a chuckle from the audience: "Maybe Don Tapscott is going to show us how to make money on the Internet." I felt myself having some kind of dot-com flashback. Sure, today there are interesting questions about how "Internet properties" like YouTube or Facebook will find a way to be profitable. But the more interesting questions are about how a gold mining company, a book publisher, a clothing manufacturer, or a retail chain can use the Internet to evolve their business models to innovate better, create new value, redesign their supply network, and engage their customers and the world in new ways.

The smartest managers already know that user-generated media and social networking are really just the tip of an iceberg. Rather, nothing less than a new mode of production is in the making.

Employees drive performance by collaborating with peers across organizational boundaries, creating what we call a *wiki workplace.* Customers become *prosumers* by getting engaged in co-creating goods and services rather than simply consuming the end product. So-called supply chains work more effectively when the risk, reward, and capability to complete major projects—including massively complex products like cars, motorcycles, and airplanes—are distributed across planetary networks of partners.

Dumb companies are banning Facebook and other social media. Smart companies understand that these tools and platforms are becoming the new operating system for business. Even ardent competitors are collaborating on path-breaking science initiatives that accelerate discovery in their industries. Indeed, as a growing number of firms see the benefits of mass collaboration, this new of way organizing will eventually displace the traditional corporation as the economy's primary engine of wealth creation.

Already, this new economic model extends beyond software, music, publishing, pharmaceuticals, and other bellwethers to virtually every part of the global economy. Today, encyclopedias, jetliners, operating systems, mutual funds, and much else are being created by teams numbering in the thousands or even millions. While some leaders fear the heaving growth of these massive online communities, this is folly. Smart firms can harness collective capability and genius to spur innovation, growth, and success.

Thanks to the Internet, companies are beginning to conceive, design, develop, and distribute products and services in profoundly new ways. The old notion that you have to attract, develop, and retain the best and brightest inside your corporate boundaries is becoming null. With costs of collaboration falling precipitously, companies can increasingly source ideas, innovations, and uniquely qualified minds from a vast global pool of talent. The end result is that the corporation may be going through the biggest change in its short history.

These leaders couldn't care less about eyeballs, clicks, and the stickiness of their Web sites. They don't ask outdated questions like "How do we monetize our content?" The focus is shifting to how a firm orchestrates capability and innovates to beat its competition and create sustainable value for shareholders.

It's in this spirit that I welcome the publication of *ANYWHERE.* Global connectivity *is* revolutionizing the way we do business. Experiences *are* now portable. The consumer *is* now connected. Consumers *are* becoming producers. And enterprises *are* becoming networks not places.

But how to exploit this new connectivity? Read on.

Don Tapscott is the coauthor of 13 books about technology in business and society, most recently Grown Up Digital: How the Net Generation Is Changing Your World. *His penultimate book (with Anthony D. Williams),* Wikinomics, *was the best-selling management book in the United States in 2007. He is chairman of the think tank nGenera Insight.*

ACKNOWLEDGMENTS

If Hillary Clinton was right—that it takes a village to raise a child—perhaps I shouldn't be surprised to find myself saying that it takes a company, and its clients, to write a book.

Anywhere could not have come into being without the challenges and insights of Yankee Group's global clientele and the powerful ideas and creativity of our research organization.

My thanks must go first to all the smart people who made time to talk with us for this book. There were over 50 interviewees, making the list too long to reproduce here, but their generosity with their time, experiences, and ideas has helped bring this book to life.

Special thanks to: Maynard Webb (LiveOps), Mike Muller (ARM), Doug Brown (Bank of America), Michiel Boreel (Sogeti), John Halamka (CareGroup), Steve Tomlin (Chumby Industries), Mark Templeton (Citrix), Christian Verstraete (Hewlett-Packard), Sriram Viswanathan (Intel), Nicholas Negroponte (One Laptop Per Child), Bob Metcalfe (Polaris Venture Partners), Steve Haber (Sony), Vanu Bose (Vanu Inc.), David Rose (Vitality), and Larry Weber (W2 Group).

And for sharing material and sources, thanks also to: Joe Mandacina (Sprint), Cor Dubois (Alcatel-Lucent), Nicco Mele (EchoDitto), Brian McManus (Amdocs), Sheila Burpee (BelAir Networks), Michael Barron (DLA Piper), Karen Lilla (IBM) and Beth Morrissey (GSMA).

Then to the Yankee Group core book team members, who somehow squeezed this project in and around their day jobs yet still made seminal contributions to content and quality: Josh Holbrook, research director, Carl Howe, research director, Yang Sallette, project manager, and Brianne Shay, graphics designer. Thanks as well to our cowriter Charlotte Evans, who waded into Anywhere for this project . . . and decided she needed a Kindle!

Beyond this core team, a number of other YGers contributed time and insight: Wally Swain, Dianne Northfield, Declan Lonergan, Chris Collins, Josh Martin, Phil Hochmuth, Camille Mendler, Andy Castonguay, Benôit Felten, Phil Marshall, Brian Partridge, Shirley Macbeth, Stefanie Joyce and Zeus Kerravala.

And thanks to the entire Yankee Group organization. I'm terrifically fortunate and proud to work with such a smart, committed, enthusiastic group. Thanks to our Board of Directors for their encouragement and to our management team for their support. Thanks of course must also go to the team members at McGraw-Hill, particularly, Donya Dickerson, for their enthusiasm for the project.

Finally. Thanks to my mother, who as an author in her own right made this seem exciting; to my husband, who made it seem possible; and to my daughter, who made it seem entirely matter-of-fact. (The dog didn't seem to notice.)

PART I

Welcome to ANYWHERE

The ANYWHERE Revolution

"The Internet gave us time freedom, then freedom of choice. Now it's giving us 'where' freedom."

—Hilmi Ozguc, founder, Maven Networks

Quick quiz: What impact has technology had on your life in the past five or ten years? The chances are good, no matter who you are, what you do, or where you live, that you'll say that there have been some profound changes to how you live and work. Here are just three technologies that I can probably guarantee you have personal experience with: the Internet, the cell phone, and the digital camera. Each one has been a game-changer, transforming what we do, how we do it, and when and where we do it.

The technologies at the core of these examples—a common digital network, wireless data transmission, and a standard digital representation for pictures—have not only changed us as people, but they have reshaped the business landscape for those activities and across those industries. Think about it: Online commerce redefined shopping forever, putting some retailers out of business and forcing the survivors to change. Cell phones have expanded from being used in our cars to every place we go, including inside our own homes, because many of us have decided we don't need a home phone line anymore. And digital cameras—when was the last time you saw a picture-developer kiosk in a mall parking lot? Polaroid went out of business, and Kodak is a dramatically different company from what it was 10 years ago. Billions of dollars of economic change have been set in motion from just these three developments. So yes, technology has brought big changes to us in the past decade.

Across the developed world in North and South America, Europe, and many parts of Asia and the Pacific Rim countries, the vast majority of us use one or more of these three technologies on a daily basis. And in emerging markets, the advance of technology has been very different but no less dynamic. Even though many parts of the world haven't yet tasted all the technology breakthroughs of the past decade, they've been experiencing breakthroughs nonetheless. Cell phones have taken off like a rocket in markets that have lacked the most basic telephone infrastructure. Indian medical experts examine digital X-rays e-mailed to them overnight to speed diagnosis and reduce health-care costs. Pirated digital copies of movies and music show up in the most remote parts of the world.

Here's the second part of our quick quiz: There's a common thread among these three technology examples. What one trait do these particular breakthroughs have in common with each other?

The answer is *connectivity*.

Each of these changes enhances our ability to connect with others. A digital connection is an element of each device's value proposition to us: the Internet's global collection of information resources, the cell phone's ability to take our calls and messages while we're on the go, and the digital camera's ease of reuse and retransmission of our images.

These developments, along with all of us, are just a few of the actors in a grand connectivity story playing out around the world. They are bringing us closer to a future of *global* connectivity, when all people, and all the things we care about, will be connected to each other on a vast digital network. In fact, the word "global" doesn't do this justice, since it fails to clarify the extent of the expansion not just to more places and more people but to more things as well. The right word is actually "ubiquitous."

What is happening, precisely? Simply put, a convergence in the advances of communications technology, the powerful economics of large scale, and our fundamental human appetites are coming together to accelerate the expansion of connectivity in our lives.

"I know already," you may be saying. "I'm plugged in, I'm connected." Probably true: many of us are attached to iPods, use **Wi-Fi** from the road, and nag our children to stop texting long enough to eat a decent breakfast. We have a reasonable chance of finding cell phone reception wherever we go. And we're getting bombarded with commercial messages of all kinds about connectivity: Change your mobile phone plan! Get your TV entertainment from your phone company instead of your cable company! Save travel expenses at work and do a videoconference!

And certainly the Internet's emergence has changed our world: We order things online. Many of us buy movies and TV episodes on iTunes or other popular Web sites. Many of us are also changing the way we get news and reading material by signing up for e-mail alerts, checking online news sites, and, most recently, using digital reading devices, or e-books, to download and read books, newspapers, and magazines. Some of us can even access the Internet on game consoles hooked to the TV.

That said, our connected experiences are uneven and mishmashed. We can access the Internet on our phones, but we can't easily edit our documents on them. We can download books, music, and videos to portable devices, but we can't readily access content other than the specific media targeted for each individual device. We use multiple devices (laptops, cell phones, TVs, iPods, e-book readers), but many of them don't connect easily with one another, causing our drawers to fill up with mysterious black cables and plugs whose purposes we quickly forget. We're straitjacketed by existing technology formats and investments, laws about intellectual property rights, and simple convention. Our emerging world of connectivity is without order—or is it an excess of order?

The truth is that we've just begun a journey that will take us to a future very different from today's sometimes rewarding, sometimes bewildering connectivity experiences.

This expansion of connectivity that we've witnessed, bringing us all three of these successful new technologies and many others, is a work in progress, building on both the successes and failures of recent connectivity offerings as well as breakthroughs in technology and price that still lie ahead.

It's been sparked by a potent trio of forces—a common digital network, the availability of broadband capacity, and the powerful economics of wireless transmission. (See Figure 1.1.)

These elements have already begun to create explosive change for all of us, even as they combine in new ways to change us even more. Here are a few global connectivity developments already on the horizon:

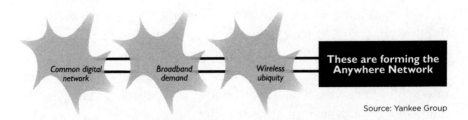

Source: Yankee Group

Figure 1.1 *Sparks that ignite the Anywhere Network.*

- **More connected devices:** Vitality, Inc. has developed a pill bottle with the capacity to notify the network when it's been opened or closed. Today, only about half of patients who have been prescribed a daily pill actually manage to stick to the regimen. The rest forget to take their prescribed dosage at least some of the time. For common conditions like high blood pressure and diabetes, lack of complete adherence to medication means that patients don't get the care they need, and health-care costs, already high, spiral even higher. Data gathered on pill consumption with the help of the container itself can help family members and medical personnel increase patient compliance with the prescription, thus improving health while reducing the expense of monitoring the activity.

- **More connected experiences:** The growing role of connectivity in our day-to-day experiences is sometimes as invisible as the radio waves that enable them. Vending machines in Japan, a very popular method of selling everything from sandwiches to dress shirts, are beginning to report on their inventory over a wireless telephone network, thus saving their operators wasted trips through traffic when the machines don't need to be replenished but ensuring that they do send a truck when the product is running low. In multiple markets including South Korea, products can be paid for at the point of purchase by passing a mobile phone in front of a wireless reader; in the Netherlands and elsewhere, parking meters have been replaced with signs indicating the process for paying for the spot by texting from the driver's mobile phone.

- **More connected businesses:** Connectivity expansion is touching much more than consumer devices and activities; businesses are at the forefront of a tremendous explosion in connectivity-enabled assets and processes. Airports in Hong Kong and Malaysia put wireless transmitters on airplane catering trolleys so that they can be rapidly and automatically routed through a stripping and reloading center to put the correct food on each unit and send it back out to the right flight. Heathrow Airport has begun to incorporate wireless transmitters on taxis to reduce queuing time for both passengers and taxis.

This connectivity expansion is unstoppable. In the next 10 years, the world's people and things will be transformed by a steady expansion in connectivity among them.

The total impact of the changes being brought about by these communications breakthroughs is magnificent in scope, and we've truly witnessed only the beginning. This remarkable set of developments will change who you are as a person and what your company should be doing to reach your customers, workers, partners, and assets.

And that's what this book is about: the changes still ahead in connectivity. They will bring us the largest technology change in our lifetimes. No one and no company will be spared as it revolutionizes how we do business. And businesses that see what's coming and know what to do about it will have a tremendous opportunity to profit from the change.

Where (or What) Is Anywhere?

Even though "ubiquitous" is the most appropriate adjective to describe the steady progress of global connectivity to a completely pervasive state, and I'll continue to use it here and there as we talk about the increasing expansion of communications around us, I have a much simpler way to talk about this world we're headed to: **Anywhere.**

That's my name for this future state: when all people, and all the things we care about, will be connected.

Why do I call it that? Because at its very core, this transformation in connectivity is about the emergence of a global network, bringing a major shift in the relevance of *place* in our lives. This transformation is changing the meaning of location in two dimensions. First, location—where we are or where other people and things are—will become less significant to our activities and decisions, thereby reducing constraints on what we do or how we do it. Want to reconnect with old friends halfway around the world? The Internet has already helped us find more ex-boyfriends and girlfriends and former classmates than we imagined possible. Can't get a flight from Tokyo, or don't want to move from rural France? Don't worry—you can still do work for the London office using the global network. Left some information at home? Never mind; storing your personal files in the network means you can get to them from wherever you are.

But simultaneously, in a thrilling contradiction, it will also *expand* the significance of place. The fizz in our Anywhere brew is wireless connectivity, cutting the cords in our lives. Our ability to use nothing more than airwaves to economically deliver connectivity to locations not reached by fixed wires—jungles, deserts, trains, oceanliners— means that the network can finally go everywhere it can be useful. People in what have been considered the most remote parts of the world will be reachable, some for the very first time. And as the **Anywhere Network**® links businesses to all locations, business executives who previously only needed to think about the markets of London, New York, and Tokyo will have to become familiar with the needs, cultures, airports, and cuisines of Mumbai, Dubai, and Dakar.

Location will thus both matter to us less, and more, than it ever has.

Not only will Anywhere be bigger than other technology transformations we've lived through before, it will also be *faster*. Mobile phones reached their first billion users faster than any other technology introduction in history. In 2007, there were 2.8 billion mobile devices in use globally. That total has been growing since then by approximately 1.6 million new mobile phones *every day* and shows little signs of stopping. (See Figure 1.2.)

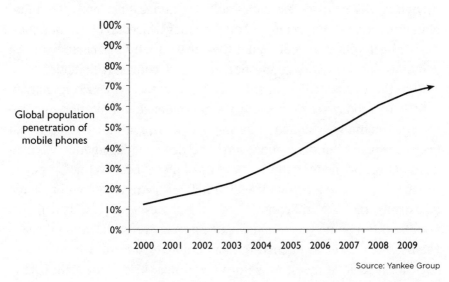

Source: Yankee Group

Figure 1.2 *Mobile phone explosion.*

The falling cost of access to the growing global network, the financial opportunity that the network's expansion is creating for firms highly motivated to benefit, and the economic power that network access puts in the hands of its users all conspire to make the move to Anywhere something like a brushfire—racing into green fields at a tremendous clip.

How Big Is Anywhere?

When most people think about connectivity, the Internet and mobile phone technology are what leap to mind. But as connectivity comes to more things around us, its contribution in our lives will expand well beyond our understanding of today's PC-based Internet and mobile phone capabilities. In the next few years, we will see an explosion of other connected devices as well. Already, we can upload photographs to the Internet without using a PC by using a Wi-Fi-enabled camera. And we can download books directly to electronic devices that store multiple titles digitally and display them page by page, subject to the reader's control.

The emergence of Anywhere is also going to affect what types of connections we have to the Internet and to each other. The network behind Anywhere will let us access data—software applications and all manner of information—"in the cloud." You'll no longer need to be at your computer to get your personal digital files or to reach your customers. All you'll need is a network connection and some connectivity device.

But this change will challenge old ideas and ways. Our connectivity experiences today are dominated by the influence of the prior communication technologies such as newspapers, television, and traditional telephone lines. Our perceptions are largely controlled, whether we realize it or not, by the inherent limitations of these older technologies.

As we move to a common global network, Anywhere will reshape us as consumers; it will change how businesses work; and it will also remake the business landscape, pitting massive multinational firms against each other in pitched battles to deliver the Anywhere experience we'll seek. Some of those firms will not survive the fight. Many new firms will emerge that pinpoint new needs stemming from our insatiable appetite to take our experiences with us wherever we go. And Anywhere will be the source of major societal change, redrawing personal relationships and norms of behavior. If that's not enough for you, political and legal landscapes will shift as well, engendering new regulation and legislation that try to support and control our activities on this pervasive digital network fabric.

I made a big claim about Anywhere earlier: that it will be the largest and most profound technology transformation of our lifetime. To make sure it's clear why we can say this, let's consider the last major technology revolution in our lifetime—the commercialization of the Internet—and see how Anywhere will compare.

Starting from the launch of the Netscape browser in 1995, the following 10-year period witnessed the growing interconnection of about 500 million PCs in the world, based in a workplace, a school, or in someone's home. For most of that period and for most people, the actual physical network connection was through the existing telephone network, using dial-up modems and phone lines to transfer bits, ever so slowly, among computers on the network. As slow and as frustrating an experience as that feels for most of us now, depending as we do on much higher-speed connections, that bit-by-bit drip of a brand-new link to the larger digital world brought about 1 billion people online. The financial value creation—the sales of computers and network services, plus the launch of brand-new commercial activities and entities dependent on the network—ran to billions of dollars worldwide by 2005.

Anywhere—the inevitable expansion of digital connectivity to all of us and the things we care about—will beat out the Internet and any other technology experience we've yet lived through. Its impact will dwarf those changes just in any single dimension alone. In the aggregate, it will produce a stunning total change that will be greater than the sum of those dimensions. (See Figure 1.3.)

For starters, the *places* connected by the network will not just be work, school, and home—and not just in the parts of the world that have been able to afford the costs of the telephone network. Instead, the network will reach into the farthest corners of the world, making the lack of connectivity a luxury rather than proof of irrelevance.

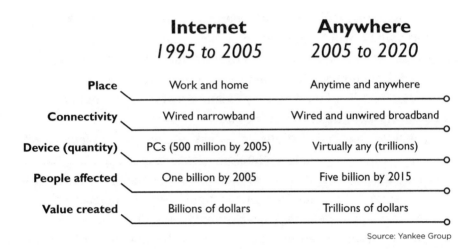

	Internet *1995 to 2005*	Anywhere *2005 to 2020*
Place	Work and home	Anytime and anywhere
Connectivity	Wired narrowband	Wired and unwired broadband
Device (quantity)	PCs (500 million by 2005)	Virtually any (trillions)
People affected	One billion by 2005	Five billion by 2015
Value created	Billions of dollars	Trillions of dollars

Source: Yankee Group

Figure 1.3 *Anywhere will surpass the Internet's impact.*

The *devices* brought to the global network through this expansion will not just be computers and phones, but cameras and picture frames, cars and crates, pumps and pill bottles: literally anything that can benefit from being able to exchange information of any kind with any other item or person in the world.

The *people* connected to the global network will be all of us—over 5 billion by 2015 just by virtue of the expected expansion in mobile phones alone. When people can send and receive voice, text, and other digital services, they've joined the global network at the center of the Anywhere era. And it's not just affecting people in the developed world. After many years of marginal access to communications technology, or none at all, emerging markets such as India and parts of Africa are finally getting connected too.

Finally, the *value creation*—the financial contributions to the global economy, creating revenue, jobs, and profit—will be in the trillions of dollars by the time the Anywhere Network has expanded around the world and has the capacity and intelligence to support an explosion of new services that will take advantage of it.

So yes: This progression toward Anywhere is big. The path to Anywhere will be a complex and sometimes bewildering drama with some exciting twists and turns ahead. In fact, not only is this going to be a dramatic set of events, I'll go yet another step and call it a revolution—the **Anywhere Revolution**.

Like other revolutions before it, battles will be waged that will have a significant impact on the ultimate state of affairs, and those outcomes will not be universally welcomed by all. And in the global landscape across which these developments play out, they will leave permanent scars and new landmarks.

The Birth of a Revolution

A revolution is a far-reaching set of changes resulting from drastic events. You can spot a revolution in several ways: a shift in the existing power structure, major economic disruption, significant changes in behavior, or global scope—all emerging from pitched battles on multiple fronts that are waged between passionate combatants who use all possible means available to prevail.

Johannes Gutenberg's printing press was one of the most significant catalysts of the European Reformation and a major technology revolution itself. From the tightly controlled social structure in place at the time, its emergence created havoc. No one was in sole control of its use, and thus it brought information to people who for centuries had been unhappy with aspects of the Catholic church. Books could suddenly be published without church authorization; a surge of secular propaganda emerged attacking its corruptions. Both the church and society at large were permanently changed by the emergence of the press. Literacy came within reach of common people; money was made by publishers selling books to an eager public; laws came into being to protect the works of authors more widely distributed than ever before. The genie was out of the bottle forever.

The Anywhere Revolution will share many of the hallmarks of the emergence of the printing press as well as other technology revolutions: it will crown new kings, shift economic benefit and harm, and leave permanent change in its wake.

The battles that will make up the revolution's core activities will be varied. Massive existing stakeholders with well-entrenched views threaten the realization of change promised by rebel upstarts. TV net-

works, mobile networks, and wire-line telephone networks were originally closed systems, their use constrained and controlled by owners,
technology limitations, and laws. Today's owners of these networks
want desperately to maintain control, defining what services are
offered, what things can connect to their network, and how the funds
flow to pay for it all. But a billion devices that can speak the common
language of the new digital network are laying siege to that model,
while Internet users accustomed to surfing to any Web site in the world
without barriers see no value in network-constructed constraints. A
morass of laws and regulations around the world will weigh in while
this struggle wears on, dictating what network providers can and cannot do with their networks, protecting the status quo or attempting to
catch up to the changes afoot. Will the winners of the TV and telephone revolutions emerge victorious in the battles shaping our
ubiquitously connected future? It's a war: all bets are off.

As with all revolutions, the changes Anywhere brings to us will be
disconcerting in one way or another. Even as we wrestle already with
connectivity change—as parents deciding if, or when, our children may
use Facebook, as managers trying to oversee increasing numbers of
employees working from home—there is much more uncertainty
ahead. But rest assured that anxiety and uncertainty in the face of technology change have been features of every major wave of technology
change the world has ever experienced.

Before the modern railway was introduced, it took days to get from
one town to the next. Carriages and horses were the principal means of
transportation, and it was impossible to transport large numbers of people at the same time in a single vehicle. When the steam-powered
locomotive emerged, suddenly it was possible to travel several hundred
miles in hours, not days. People could travel together in large groups at
an affordable price. Businesses were able to move goods from place to
place economically, opening up new markets literally overnight.

The change was monumental—but it was not seen in an entirely
positive light. From their earliest days, railways inspired deep anxieties
among the general population. Many people were entirely opposed to
them, believing they were fundamentally unnatural. Historian Ralph
Harrington explains that people thought "trains would blight crops

with their smoke and terrify livestock with their noise, that people would asphyxiate if carried at speeds of more than 20 miles per hour, and that hundreds would die yearly beneath locomotive wheels or in fires and boiler explosions."

Beyond worries about physical harm, the railway was seen by many as a threat to the social order. Many thought it was far from a liberation; rather it allowed the lower classes to travel *too* freely, potentially weakening moral standards and dissolving the traditional bonds of community.

As laughable as some of those concerns appear today, history repeats itself. Among connected people in today's developed markets, we already worry about the potential negative affects of connectivity in our lives. Do mobile phones make it easier for children to cheat in school, to disengage from the fabric of their families, to fall prey to unknown predators? Are remote workers really working when you can't see their desks? Will connected displays and instantaneous information turn us into a herd of mush-minded, media-addicted cattle who lose the ability to think for themselves? Tangled up in eminently reasonable questions—about how this transformation will change our lives and how the values we hold can still be respected—lurk unreasonable fears, too. They will subside only with the progression of time and the gradual acceptance of change.

Anywhere as a revolution? Yes, it ticks all the boxes: the threat of a change in who rules the connectivity fabric we'll depend on, the shift in economic structures to change how and where companies make money, the introduction of permanent new behaviors—all at a global scope.

There's Gold in Them Hills

In the late nineteenth century, as railways began to snake their way across the broad expanses of North America and Europe connecting distant towns and hamlets that previously had only sporadic contact with even neighboring metropolises, businesspeople with acumen seized the commercial opportunities they presented. One was a railroad station agent in Minnesota in 1892 named Richard Sears. He saw the potential to use the capacity and dependability of the new North American rail system to deliver consumer goods to farmers and others who couldn't easily reach urban stores and marketplaces. Ultimately his vision helped build Sears, Roebuck, a mail-order retail business that in only three short years after its inception had a 500-page catalog of goods not readily available to a rural population—all delivered by rail. That catalog became a window to the world for the firm's customers, and it changed their lives and aspirations.

The emergence of the Internet created a similar rush to build profitable businesses premised on its connections with consumers and businesses. Excitement about its potential skyrocketed in the mid-1990s. As hype bred more hype, eventually in 2000 the so-called "dot-com crash" destroyed the dreams of many start-ups and their investors, and brought the market valuations of many surviving Internet firms back to earth. Yet in that wake were many firms that could not have existed prior to the Internet, still with substantial valuations in the open market. Even after the tech crash of 2000, Amazon's market capitalization was still over $15 billion, and literally thousands of firms around the world had begun to remake their businesses to take advantage of a new, brilliantly simple and fantastically powerful way to reach customers and partners. The total commercial value of the Internet's emergence in a 10-year period was easily in the hundreds of billions of dollars.

The expansion of the global network toward Anywhere will create similarly exciting opportunities, both for the firms that help build out the network itself and for those that deliver products and services that take advantage in compelling ways of the reach and capabilities of that digital fabric.

I lead Yankee Group, a company that researches and predicts changes in global communications technology. We put the total potential value creation from Anywhere over the next 10 years in the *tens of trillions of dollars*. That value creation will come in two main areas—the network itself, providing the platform for change, and the things and services that use the network, riding atop its connections like so many sewing machines and bags of seed loaded onto rail cars bound for distant farms.

We estimate that the expansion of the Anywhere Network itself will generate almost $1 trillion in global revenues just by 2012. Those are proceeds that will go to the builders and operators of that global fabric—certainly, some current firms that provide network components and services today, but also some new firms that may still be just a gleam in an entrepreneur's eye. (See Figure 1.4.)

We arrive at this figure by examining the **Anywhere Network Economy**. That's what we call the annual revenues generated by business and consumer broadband connectivity around the world, coupled with the network hardware and software acquisitions made each year to help deliver those connections. Consumer spending dominates the global Anywhere Network Economy in 2010, accounting for $566 billion of the $754 billion spend. This shouldn't be surprising, since that $566 billion is supporting more than a billion wired and wireless broadband users worldwide; even with a relatively low average revenue per customer, the sheer number of worldwide consumers makes this the bulk of Anywhere Network revenue.

We consider our Anywhere Network Economy numbers to be a bare minimum of the true amounts that will be spent. That's because these projections don't include the revenues from the sale of the devices that access the network—PCs, phones, and the many other kinds of products that will become connected—and they don't account for the wages earned by people running the network. So the true baseline economy may be much larger—but we can be certain that it is not smaller than the estimated annual global spend presented here.

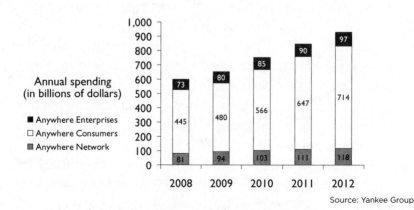

Figure 1.4 *Anywhere Network Economy revenue expansion.*

Growing atop that economic platform will be a tremendous flowering of new devices and services that depend on its capacity and capabilities. The sheer breadth and variety in the types of things that will connect to the network and the services that will be provided, to say nothing of how they are used to create financial benefit for their providers, make this second component of Anywhere's commercial value creation fiendishly difficult to estimate and leave us searching for useful parallels.

The development over several decades of the U.S. national highway infrastructure—numbered and signed routes that allowed drivers to travel to the four corners of the country on a single system of roads—spawned many new businesses. Motels were one such new industry, sprouting up along those routes to give tired travelers an easy place to stop for the night.

In 2006, the U.S. Transportation Research Board produced a summary of a number of seminal research efforts to estimate the total value creation that emerged from that national commitment. It identified three broad areas of net positive contributions for new highway infrastructure: reduced industrial costs, increased productivity, and net rate of social return. Between 1954, when the U.S. highway build-out was well under way, and 1974, real GDP in the country went from $2 billion to $4 billion—it doubled.

With the emergence of Anywhere, new goods that incorporate connectivity will be created and sold. Connected consumer electronics, connected health-care devices, connected cars, connected industrial assets, and much more will be part of this development. Just these new versions of existing products, along with entirely new product categories of items such as personal navigation devices (PNDs), could represent trillions of dollars in annual revenue by 2015.

Connected services—media, information, money, and all the activities that can take place in a pervasively connected global society, delivered to us anytime and anywhere we want it, on any device that's capable of using it—represent the other broad area of value creation atop an Anywhere Network. Will this all be new revenue? Hard to say. That's one of the things that make estimating the total financial impact of a global network so hard.

Some services will be entirely new activities. I may choose to pay a firm to help me monitor the security of my data stored online, backing it up periodically, sending me reports on how frequently it's accessed, providing me with some secure means to unlock it or share it with the people and businesses that I select. That's not a connected version of anything that I spend my money on today.

But many more services will be transformations of activities we do today by more traditional means. Those services could ultimately replace the vast majority of those industries' current revenues from more traditional sources. For example: The total value of advertising revenues in the U.S. newspaper industry in 2008 was estimated by the Newspaper Association of America to be $35 billion. By 2015, as more newspapers fold and revenues of news-gathering organizations shift to either advertising or fees paid from delivering their services via the network to connected devices, that figure may represent the majority of funds emerging from news gathering.

Trillions of dollars in revenues from network provision, and trillions more in connected goods and services, whether new or shifted: This truly will be the largest technology revolution of our lifetime.

Welcome to Anywhere

Billions of people? Tens of trillions of dollars? Those are big numbers to be tossing around. Given that I'm talking about dramatic stuff here—a game-changing transformation of global proportions, the biggest technology change in our lifetime—you're entitled to wonder how this book is going to deliver on its subtitle: *How Global Connectivity Is Revolutionizing the Way We Do Business.*

At Yankee Group, we're connectivity experts. For over 40 years we've helped businesses around the world profit from the changes we foresee. In a way, we're the tornado chasers of connectivity change. We gather data to define the trends we're seeing and then use those data to predict the coming weather. We help our clients to anticipate the future and make profitable plans as a result.

In just the past year, we advised companies that build communications networks all over the world on what businesses and consumers

Anywhere Thinkers on the Anywhere Revolution

How big is the global impact of the Anywhere Revolution—the emergence of ubiquitous connectivity? So big that it can be challenging to describe the scope adequately in a few words. Here are some dimensions of impact from some of the connectivity industry's leading thinkers.

◗ Social and Economic Implications of a Connected World

"The global network will provide unbelievable connectivity and intelligence. It might be the most exciting development in the world. It can bring nonlinear change in every economy."

—Dov Baharav, CEO, Amdocs

"For the poor, being connected helps break the cycle of inefficiency and exploitation. Even factoring in the cost of connectivity, you can spend less on credit, water, all the other things you need."

—Rajeev Suri, CEO, Nokia Siemens Networks

"When mobile phones began to take off in Latin America, I saw a sign in the street one day. It said, 'I'm José, I'm a plumber, and I have a cell phone.' So I called him. He told me his phone transformed his ability to get work."

—José Maria Alvarez-Pallete López, CEO, Telefonica International

"Connectivity completely changes what illnesses are treatable. We can find patients, we can build information about them, we can provide continuity in their care. And when patients think the health system can help them, it changes their perception of what's possible."

—Dr. Hamish Fraser, Director of Telemedicine, Partners in Health

"Where we connect children with networked computers, school truancy drops to zero. Connected children are teaching their parents to read and write, too."

—Dr. Nicholas Negroponte, Founder and Chairman, One Laptop Per Child

◗ Changes We Can Expect in the Anywhere Revolution

"Connectivity will become multimegabit and ubiquitous. It will be a change as profound as the personal computer."

—Sriram Viswanathan, Vice President, Architecture Group, Intel

"Two massive changes are taking place. First, pictures are taking over from words. The network allows people to use pictures and video, and we want to. This will have a huge impact on how we communicate. The second is about expectations. The generation after me expects the same experience wherever they are."

—Ben Verwaayen, CEO, Alcatel-Lucent

"We have an insatiable demand for connected services. We expect more. We'll see tighter integration between consumer activities going forward, with less of a distinction between, for instance, banking and buying things."

—Doug Brown, Senior Vice President for
Mobile Product Development, Bank of America

"We'll be surrounded by devices but we'll be in control. It will change the way we experience any kind of media. We'll build, evolve, add on and change them."

—Larry Weber, Chairman and CEO, W2 Group

▶ Commercial Implications of a Comprehensive and Capacious Global Network

"Enterprise computing, the way it works and what people see in an office today, is on the way out."

—Mark Templeton, CEO, Citrix

"Vast computing resources will be summoned from the network with zero capital expense."

—Marc Benioff, Chairman and CEO, Salesforce.com

"Smaller enterprises will be on a more equal footing with larger ones; they won't need 200 IT people to use state-of-the-art tools."

—Terry Stepien, President, Sybase iAnywhere

"We'll have our workforce in the cloud along with our assets—completely changing the way we manage work."

—Maynard Webb, Chairman and CEO, LiveOps

want from their networks. We worked with a major global retailer to figure out how to sell more mobile phones to its customers, looking at the kinds of customers it serves and matching up customers' appetites for phones with specific marketing approaches. We assessed the opportunities for a credit-card company to partner with mobile phone network operators to let consumers make credit payments by mobile phone. We helped enterprises determine how to use the global network to lessen their dependency on internal information technologies and to help their employees be more effective. In every case our focus was on how to assess the state of expanding connectivity and match that up with a commercial strategy to profit from it.

And that's what I want to help you do too: I want you to profit from the Anywhere Revolution.

To do that, this book is organized in four parts. The first is about the Anywhere Revolution itself—what it is, why it's happening, and where and how quickly it is bringing change around the world.

The second part is about us as consumers. The advent of portable, connected devices is turning us into a new kind of consumer, gradually expecting to take *all* our experiences with us anywhere we go. What other devices will take on connectivity and why? What new kinds of experiences will we seek? How will this change us as consumers, and how as businesspeople can we take advantage of the opportunities? We'll take a trip to the future in some interesting places around the world, and then return to understanding how the emergence of Anywhere is already beginning to change us and how firms should market to us.

In the third part, we look at the impact of Anywhere on enterprises. As we each become an **Anywhere Consumer**®, we'll begin to expect constant continuity in our activities with businesses—as customers, workers, and partners. Adding connectivity to important assets in business will help us collapse the costs of time and distance. And as a capacious global network allows our businesses' internal information technology infrastructures to be simplified, there will be enormous profit opportunities in the restructuring of enterprises as we know them today into very different **Anywhere Enterprises**®.

Having visited the impact of Anywhere on us as individuals and as businesspeople, the fourth and final part of the book zeroes in on how to think about profiting from Anywhere—when to move, how to move, and what to be aware of as you do it. You'll find ideas on how to win more customers, keep them longer, sell them more things, and make those things with less expense.

Throughout the book, I show you some of the data that Yankee Group collects to help our clients. Each year we survey thousands of consumers and enterprises about their connectivity experiences, and we speak with hundreds of connectivity technology executives about the direction that connectivity solutions are taking. This means that we can make some pretty smart guesses about the transformation and track its progress with figures and stats.

"Hmm: technology and data," you may be thinking. "Maybe this isn't for me." Don't worry. This isn't a book about technology per se; it's about the *impact* of technology and how it is revolutionizing business. We keep the techie stuff to a minimum; any terms that you might not know are **bolded** the first time so you can look them up in the glossary at the back of the book. As for the data: I show you just enough to prove my points; the notes section can help you find more if you need it.

And to help you visualize the impact and how to benefit, I bring some other experts into the conversation. More than 50 thought leaders in the Anywhere Revolution—executives and other pioneers in developing and using connectivity—have been kind enough to speak with me specifically about the questions this book will answer for you. You've heard some of them talking about the impact of Anywhere in the sidebar on pages 20–21, and you'll hear more from them throughout the book, sharing experiences they've had and lessons they've learned.

Deeper interviews with some of these thinkers can be found on a Web site associated with this book (anywhere.yankeegroup.com), along with lots of other resources to help you proceed.

* * *

In the past few years, the global business climate in which we all live and work changed dramatically. News headlines described rapidly worsening economic conditions, frightening many of us and our leaders. Under the strain, many businesses around the world flagged; some failed. At the same time, it left consumers fearful and unwilling to spend. None of us escaped some form of impact.

All technological changes can carry the world's good or evil along with them. But Anywhere—a powerful revolution of technological and economic forces bringing digital connectivity to all of us and the things we care about—is the printing press, the railroad, the electrification of our cities and towns, the introduction of radio and television, and much more. It's a set of inexorable changes propelling us forward with tremendous upset and upheaval, on a largely positive path that will lead to growth and exciting new commercial opportunities. It's time to understand what's happening and what you can do to stay ahead of the game. So let's get started. Welcome to *Anywhere*.

Chapter 2

Going from Somewhere to ANYWHERE

"This revolution can't be stopped. Do we stop building airplanes because they can go places we don't like? People want to be connected, and they will get it one way or the other. They always do."

—Ben Verwaayen, CEO, Alcatel-Lucent

Anywhere—the emergence of **ubiquitous connectivity**, connecting all of us and the things we care about—is a worldwide revolution, one that will change entire societies and move billions of dollars across borders during just the next five years. But how did we get here?

The purpose of this chapter is to identify the ingredients of this revolution and to understand how they're coming together to push us forward into this new world. The goal is to eliminate any lingering doubt in your mind about the nature, power, and inevitability of these changes so that you can then start the job of determining how to take advantage of them in your business.

Let's begin by ensuring that we agree on what this revolution is *not*.

It is not just a fancy name for the ongoing maturation of the Internet. This now indispensable worldwide digital platform for business, entertainment, information, and more would not have emerged on the scene in the 1990s without the first element in our Anywhere Revolution. And the Internet continues to evolve today. But Anywhere is more than the next cool social networking site or the admittedly impressive expansion of Google's aggregation of digital assets.

And Anywhere is not just about the growth of mobile phones, either. While wireless technology is an essential element of this transformation, Anywhere is about every phone, PC, laptop, and BlackBerry on the planet today, but it's also about many, many more items in our lives that will ultimately share some of the same connectivity abilities that these devices have today. Items that we might find hard to picture with connectivity, like our microwaves, cars, and even office equipment.

No, the Anywhere Revolution is more than the maturation of the Internet and more than the explosion of mobile phones. The emergence of ubiquitous connectivity embraces the Internet and embraces mobile phone technology, but it drives these innovations forward into an exponentially more powerful whole that will ultimately make both these developments look as primitive as first-generation cars, telephones and other technology innovations eventually do. It's about how these devices function, their capacities and limitations, and their potential. It's about the technology behind their connectivity, and it's about our own human interests, needs, and desires that drive our progression toward a state of constant contact.

To understand why and how the resulting communications fabric may ultimately look, let's first explore the three most essential ingredients of the Anywhere Revolution: convergence toward a common digital network, the world's insatiable demand for increasing capacity in the network, and the game-changing economics of wireless transmission. If you're going to share my conviction that these elements are going to be the key source of the largest technology transformation of our lifetime and that how you do business will be revolutionized, then it's important to take a few moments to understand them and how they're combining to create such motive force.

The Three Ingredients of Anywhere

Revolutions in history are extraordinarily complex chains of interrelated events—melding at a bare minimum social, military, and economic factors. Academic careers are made and lost over decades of painstaking research and debate as researchers attempt to tease out the critical actors and seminal events that led to massive change. One challenge in any analysis of a revolution is where to start. Did the French Revolution begin because of an extensive famine among its poorer citizens? Or the unusual weather that caused their crops to fail? Or was it the choices of what to grow?

I suppose if we were going to be exhaustive in itemizing the innovations and developments leading up to this Anywhere Revolution, the emergence of ubiquitous connectivity, we could start with the transistor. After all, it is the one common element in televisions, radios, computers, and phones—the first completely ubiquitous component in modern electronics, and essential to our progression to a ubiquitous global network.

From there we could chart the progress of computing, digital storage, and other indispensable engineering developments of the twentieth century without which we'd have neither a global network nor any way to take advantage of one.

But let's skip the origins of eighteenth-century famine in France, so to speak, and focus instead on the three most immediate and essential elements of what will make this latest technology shift the largest we have yet witnessed.

All IP, All the Time: A Common Digital Network

The first ingredient of Anywhere is a *common digital communications network*, one that is capable of carrying anything that can be reduced or translated into binary bits: voice, music, and images of course, but also money and oceans of information so vast in size and variety, and growing so quickly, that they're virtually impossible to characterize.

Today's commercial Internet burst into mainstream awareness in the mid-1990s. But the core idea behind it—breaking up digital data into packets that can navigate differing networks to their final destination and be correctly reassembled on arrival—started much earlier, with the development of the **TCP/IP suite** in the early 1970s.

Since telecommunications began, information has been transmitted over a range of independent, sometimes unconnected, networks that vary in their technology, each one originally purpose-built for specific jobs with unique features related to the task. **Fixed-line** telephone networks—first composed of cabling running across the landscape on tall poles, connecting homes and offices with telephone-company-owned switching equipment—were built expressly to handle voice calls alone, even supplying their own power to do so in the event that the electrical grid might fail. Cable TV networks emerged in the mid-twentieth century. They used an entirely different type of copper wire to carry analog TV programming to homes that couldn't receive it well through broadcast means. If you had no power in your home, you wouldn't be watching TV, so no such power solution accompanied that network. For a time these independent and unrelated networks coexisted quite peacefully, their owners never suspecting that one day they would be competing with one another.

Then the emergence of commercial computing began to create vast new libraries of digital information, and caused us to begin to develop ways to transform traditional media into digital form. With the creation of the Internet protocol suite in the 1970s, **IP** emerged as, essentially, a common language for reliably conveying any form of digital data over any type of network that can speak it. The development of modems, which first adapted the analog telephone network for the transmission of digital data, began a wave of use of a stand-

alone network for something other than its original purpose. Accelerated by the rise of dial-up Internet services in the 1990s, both telephone and cable TV networks began to be used more and more for data transmission. Their owners suddenly found themselves in competition for the exciting new opportunities they all saw to increase the value of their networks by adding new things to charge their users for. (See Figure 2.1.)

Do we need both a telephone network and a cable network in any one location when both can do essentially the same job? With massive amounts of information now in digital form and a common language to distribute it, major ramifications are playing out in the communications sector and in legal and regulatory contexts today. The simple reason is leverage: When one digital network can support any conceivable form of communications, the value of single-purpose networks for voice, television, or anything else plummets.

Leverage: the good news about IP for network owners was more ways to use the network, thus more sources of revenue. The bad news: new sources of competition from other single-purpose networks that hadn't previously been a threat. This convergence toward multipur-

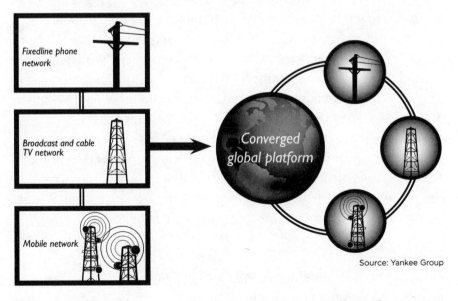

Source: Yankee Group

Figure 2.1 *Proprietary networks converge.*

pose networks has meant a worldwide scramble of owners to rapidly move to being as much, or more, multipurpose than their competitors.

Broadband Demand

The second element in our Anywhere Revolution concerns the capacity of a network, which is usually referred to as its **bandwidth** and expressed as a rate: some number of bits transferrable in one second. Think about bandwidth as you do the number of lanes on a road. Without the capacity to carry all the digital packets that want to use it, an IP network is about as useful as an alley in a small Italian hill town. A few Fiats can squeeze through, but tour buses? Forget it.

How much bandwidth does a network require? Tom Watson, Jr., the son of the founder of IBM, famously estimated in 1958 that the worldwide market for computers might extend to as many as five. It now seems laughably naive, when there are more than half a billion PCs alone in use around the world, and the true total of all types of computers in use is effectively impossible to know. Similarly for many years in the data networking business, it was common sport to argue about how much capacity a network ultimately would require; such discussions were rife with debaters who struggled to imagine the need for more capacity than was available at the time. In the mid-1980s, I was a member of an industrywide committee to develop the technical specifications for a way to transmit data at rates of up to 100 **megabits per second (mbps)** on **optical fiber**—at a time when the fastest networks in offices and corporate campuses operated at no more than a tenth of that speed. Our efforts were ridiculed by those who could not imagine the demand for such capacity. Today, corporate office networking equipment frequently transmits data at a **gigabit per second (gbps)** or more. Network capacity debates have diminished, as many in the communications industry understand that there will always be an insatiable and unstoppable demand for more and more network capacity.

Why? The short answer is this: pictures. One may be worth a thousand words, but they require a lot more than a thousand bits. Media of all kinds, including music, movies, and high-definition television can now be rendered into useful digital form, but they require orders of magnitude more bits than their simpler digital cousins of e-mail and basic Web pages. As digital media begin to swirl farther around the world, networks are scrambling to keep up. Paul Sagan is the president and CEO of Internet service provider Akamai Technologies, which accelerates the delivery of Web content and the performance of online applications. His firm has been a central player in new types of demanding network traffic. "We're starting to see the Internet effectively become television. Call it the 'HD Web,' if you will. Last year Akamai supported the U.S. NCAA college basketball championships by transmitting live-game video over the Internet for CBS, as we have for many years. In past years, we had served audiences on the Internet that were as large as some cable TV program audiences, but for the first time, a majority of viewers of the tournament selected HD video streams. That's a stunning milestone."

Couple the explosion of video and other forms of media with the ballooning number of connections to the global network—both in the expansion of individuals connecting to the network and in the number of devices each of us uses that touch the network—and you have three parallel growth exponents: more people, more activities, and more things.

The final aspect of demand is the number of simultaneous sources and targets for information transmission. When there is one source—a TV broadcast for example—there are ways to use technology to limit the amount of capacity the network needs for sending that same transmission to large numbers of recipients. But the advent of "many-to-many" networks—when all of us can be both source and target for baby pictures, home movies, X-rays, and more— places further demands on the network's capacity for simultaneous transmission.

Taken together, these factors are unleashing an unprecedented hunger for network capacity. There's an insatiable, almost infinite demand ahead for **broadband**: network capacity of at least 384 **kilobits per second (kbps)**. Before those of you fluent with network capacities jump up at this point to challenge where I've set the bar, I'll acknowledge that the precise level of network capacity implied by the word "broadband" is debatable. In North America and Western Europe, broadband connectivity tends today to mean anywhere from 1.5 mbps to 10 mbps on a wireline connection, and anything over 250 kbps on a wireless connection. "But in Japan," points out Dr. Botaro Hirosaki, an executive with leading electronics manufacturer NEC, "people already enjoy very high-speed services relative to the United States and other markets. There, broadband is at least 20 mbps and as much as 100." We return to this discussion of what broadband is in Chapter 3, since it will help us decide when the ongoing network expansion around the world will affect various regional markets.

One thing isn't debatable: Even if we could agree on a definition today, it would change tomorrow, just as our concept of TV picture quality advances as we adjust first to DVD quality, then the first HD images, and now the more advanced screens. "The definition of broadband won't be static," agrees Walter B. McCormick, Jr., president and CEO of the United States Telecom Association. "It will largely be in the hands of marketing folks, though. We'll see terms like 'Broadband Plus,' 'Ultra-Super Broadband,' maybe even 'Hyperband,' all aimed at connoting the quality of service that companies will want to offer to network users."

What do you really need to know about broadband for our purposes? Just these two core points: First, that demand for network capacity will continue to rise exponentially through the next decade at least; and second, that network operators—to the extent that they want to be competitive in the Anywhere Revolution—must continue to renovate and expand their networks to follow suit. It's not going to happen overnight, and it's not without cost, but technology and the laws of supply and demand will ensure that it does.

Wireless Ubiquity

Along with our first two Anywhere ingredients—a common network language to carry anything anywhere there's a network and the insatiable demand driving network capacity forward—a final catalyst sparked the revolution that's now under way: *wireless ubiquity*.

The reason is simple: Air costs less than copper or glass.

Wireless telephone networks evolved, of course, to enable us to take our phone calls on the road. Their explosive growth over the last 20 years has come in part from an important distinction in technology over fixed-line phone networks. It is significantly less expensive to put into place than those telephone wires strung from pole to pole or buried underground. The World Bank estimates that connecting an individual to a mobile network can cost as little as one-tenth of the cost of providing a new fixed-line connection.

There are still billions of people in the world who do not own or have easy access to a phone. The tremendous expense in stringing fixed-line telephone networks to where they are has exceeded the revenues that could be reaped from the investment since they have been less able to pay the costs. Large regions in Asia, Africa, and Latin America have not yet had conventional telephone infrastructure built out to all their citizens, leaving them in industry parlance to be "green fields" awaiting their first communications infrastructure of any kind. As a result, the rise in the last 10 years of the Internet, which many of us first used via a dial-up telephone network connection, has so far also passed those people by. Now, thanks to the more attractive economics of wireless phone networks, those people are gaining access to cell phones at a rapidly increasing rate. Their access to the global network will be much more sophisticated than the first phones that connected people in more developed markets, supporting not just voice but text and multimedia. The cost of connectivity for those consumers takes a much larger bite from their income than it does in the developed world, but it's a significant measure of its value to them that they make room for it in their spending. (See Figure 2.2.)

In the developed world, a variety of wireless technologies each suited to differing situations, including Wi-Fi, **Bluetooth**, **Zigbee**, and more, are extending the ability of wired networks to keep us connected anywhere we are—inside our homes, in moving transportation like planes, trains, and automobiles, and public spaces both indoors and out.

Metropolitan areas including Toronto, London, and New York are using common technology but varying business models to offer high-performance Wi-Fi "mesh" networks which provide broad outdoor connectivity for broadband-hungry citizens and their devices. In Toronto, cable operator Cogeco provides fee-based Wi-Fi access across six square miles of the downtown area. In London's financial district, **hot-spot** provider The Cloud provides fee- and subscription-access as well as wholesale access through arrangements with other operators.

Wireless spend as a percentage of income

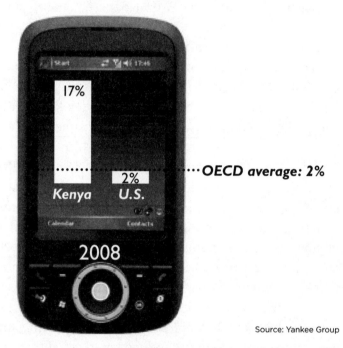

Source: Yankee Group

Figure 2.2 *Connectivity is indispensable.*

And in New York, cable network operator Cablevision has equipped high-traffic areas like train station platforms with the capability, providing it free to its cable customers.

As wireless networks continue their march to newly connected citizens in Africa and elsewhere and as the wireless networks already in use in more developed parts of the world evolve to do more for us, wireless connectivity will become the norm of the Anywhere Revolution. It won't matter whether you are in a major city or in the darkest corner of the rain forest, Anywhere will allow you to access a common broadband network.

The Emergence of the Anywhere Network

These three key factors are combining to form a global communications fabric—an Anywhere Network®. They are moving us from a network that appears in some places, at some times, for some people, to a network that is anywhere and everywhere.

What will this network be like? What will it do for us? Almost by definition, this infrastructure will be high in capacity and pervasive. But there are three additional critical qualities that Yankee Group experts expect to emerge along with this ubiquitous network, all of which are important to businesspeople hoping to build profitable endeavors that use it.

1. The Anywhere Network Will Be Seamless

Today our experiences with multiple networks are largely discrete. When we have to leave the house while we're in the middle of a phone call on our home phone line, we have to end it and resume it manually on our cell phones. Our in-home Wi-Fi network probably doesn't know anything about our home phone service. Even "**bundling**," an increasingly popular offer from network operators to provide two or more communications services in one package, such as voice and TV or TV and Internet access, is largely a superficial packaging ploy rather than the functional integration of separate networks.

But over time, we will see the gradual integration of disparate networks to work together, to provide at least what feels to the users of these networks to be a seamless experience. It's begun already in office environments, where the concept of **fixed/mobile convergence**, or **FMC**, is allowing companies to merge their corporate telephone systems to reduce expenses and the duplication of phones (wired desk sets and mobile handsets). Recently a company called TerreStar sent a commercial satellite into space to support wireless phone calls where there's no wireless coverage on the ground; the phone handsets it will offer support both existing wireless networks and the satellite system so that callers can move from one to the other. But **seamlessness** doesn't apply just to phone systems; over time all kinds of network traffic should be able to move from one type of network to another without interrupting the user's experience.

While many of the citizens who join the Anywhere Network for the first time in the years ahead will depend on an exclusively mobile connection, the ideal **Anywhere Experience** will be one that combines the higher performance of fixed broadband networks with the ubiquity of wireless, allowing us to move around our world without regard to the physical manifestation of the network that supports us in any one location. (See Figure 2.3.)

2. The Anywhere Network Will Be Secure

To carry the world's digital activities everywhere we need them to go, the Anywhere Network needs to protect its contents, whether they're money, details of sensitive negotiations, commercial secrets like patents and inventions, or just very personal information like medical test results. Today, networks vary in the level of security they provide to the traffic they carry. In the financial services sector, because of the sensitivity of the billions of dollars in electronic funds transfers that take place every day, security systems are highly advanced. On the open Internet, today's security is more limited, and the energies of thousands of miscreants around the world are focused on finding ways to steal important traffic. Network security is one of the most active areas

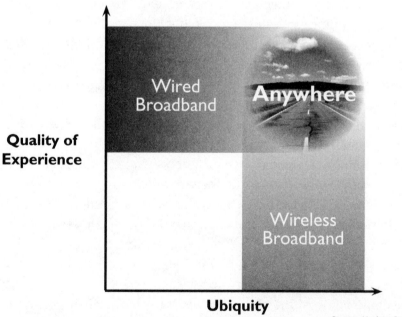

Source: Yankee Group

Figure 2.3 *The Anywhere Experience.*

of technology development in this decade, as companies work to create more robust solutions that are simpler to use and that conform to international standards.

3. Finally, the Anywhere Network Will Be Intelligent

Operators of today's networks worry about the possibility of them becoming, in the new ubiquitous network era, "dumb pipes"—by which they mean a highly commoditized infrastructure that adds little to no value to the traffic it carries, as with today's electrical grid, for example. But embedded into today's phone and TV networks are very valuable—albeit sensitive—pieces of information about their activities. Wireless phone networks know where we are; otherwise how could they send our calls to the cell tower nearest to us? And they know who to bill for the call. TV networks know what movie we ordered from their video-on-demand offerings, and again, who should get the additional charge for that service.

When this information is surfaced from the network's bowels and provided effectively to services that use the network, operators can indeed create "smart pipes"—a network with the intelligence to do more than just carry bits from point A to B. An extraordinarily fertile area of technology development now and for the foreseeable future is network software systems—technology in the guts of worldwide networks that makes it easier to build, consolidate, transfer, and use **network intelligence**. These developments will make it possible for the smarts the networks already have, scattered in various places and only available previously to their operators, to be made available for others to incorporate into more services. Ultimately, as savvy connectivity users, we may choose our network access not based on pervasiveness or reliability—the heart of many network ad themes today—but on the network's intelligence. Some thoughtful operator will be the first to stake out new marketing territory as the "High IQ Network."

* * *

So a global fabric, an Anywhere Network—formed by a common digital language, the certainty of broadband demand, and the ubiquity of wireless technologies—is evolving, which will also gradually become seamless, secure, and smart.

Where the emerging Anywhere Network has begun to touch us, it has already had a profound effect. Ubiquitous mobile services and rich broadband connectivity have enabled unprecedented levels of interaction between individuals and within communities. These technologies have also introduced consumers to a rich new world of media offerings and have begun to transform the ways that businesses work with their partners and customers. This early form of Anywhere has touched billions of individuals around the globe and enhanced their lives, but this is just the beginning. Our appetites have been whetted. We want more.

What will the Anywhere Network itself engender? I haven't called Anywhere a revolution for nothing; at the very least you can be sure that the next decade will be messy, with debate, chaos, and upheaval coming along for the ride. It can't be otherwise, and we discuss a few of the most relevant areas of concern later in the book.

But most importantly, this new way of building and thinking about networks will lead to a tremendous expansion of Anywhere products and services over the next 10 years at least. A successful technology acts like a vortex: a force that draws things into it, creating a draft, like a whirlpool or a tornado. Some of these Anywhere offerings will be continuations of ones we know already—such as smarter mobile hand sets that can do more things with a higher-capacity mobile phone network. Some of these will be new to various regions of the world as the Anywhere Network begins to reach them. And some will be altogether new and next to impossible to conceive of in advance: never-before-seen combinations of activities with connectivity, exploring and expanding on new capabilities that emerge, testing acceptance, building on the success of others, in a heady virtuous circle.

Whither Anywhere?

Throughout this book, I primarily use the word "network" in the singular. It's worth noting for purists that I don't intend to imply that this one converged network will have a single owner or operator, or even a single set of characteristics that define it equally well throughout its extent.

Indeed, the Anywhere Network will not happen in the same ways, at the same time, everywhere in the world. The starting point isn't the same—cable TV networks for instance are less commonplace in Latin America than in Western Europe. Our appetites for the network, while driven by the same fundamental needs, don't manifest themselves uniformly. You want the Manchester United UEFA cup scores, someone else wants to watch a clip of the latest Bollywood movie. And the regulatory and political landscapes that govern much of the use of networks aren't the same everywhere. Sovereign entities have different views on the value and rightful owners of the network.

So in theory, a global network; but in practice, lots of differences. That's a more complex issue we look at toward the end of the book. For now, it's a bit of poetic license that's okay for our purposes. That's because this accelerating convergence, this evolution to a worldwide Anywhere Network, requires businesses everywhere to have a plan to profit from its appearance wherever and however it happens. Your energies must be focused on recognizing the inexorableness of the change and understanding the pace at which our move to Anywhere is affecting your business—your customers, products, and processes—so you can capitalize on those changes.

Now that we have an idea of where Anywhere is coming from and why it's inevitable, let's move on to looking at the pace at which this global revolution will spread around the world.

ANYWHERE—
Where and When?

"Communications will
become global in the next
ten years. Not because
governments legislate it,
but because economics
will produce that result."

—Reed Hundt, former chairman, U.S. FCC

Your opportunity to profit from the emergence of ubiquitous connectivity—what I call Anywhere—is dependent on the evolution of a comprehensive, intelligent, seamless, digital network with broadband capacity and wireless ubiquity. And this network, emerging already, will continue to mature; of this there's little doubt. Worldwide demand, commercial opportunity, and powerful economies of scale will ensure that it does.

As businesses capitalize on the network's expanding speed, efficiency, and reach, we will see the flowering of new connectivity products and services around the world. In the parts of the world where this network is more evolved—present in more homes and businesses, with affordable capacity for people who are equipped to use it—we've already begun to see its contributions. Connectivity-centric activities like online shopping, banking, entertainment, and business-to-business endeavors already feel quite normal to many of us. Beyond that first phase of added value to the Anywhere Network, we're now beginning to see follow-on connectivity offerings as a result of the growing role of wireless in our lives. Connectivity is expanding to consumer devices such as game consoles, and wireless capabilities are coming to our cars to provide real-time navigation onboard. In business, we're beginning to recognize the cost savings in moving many conventional computing activities away from in-house proprietary equipment, to reside "in the cloud"—which is to say, using the expansiveness and reliability of a pervasive network to deliver our software and data to us on demand, rather than forcing it to be physically located wherever we happen to be.

And as the tremendous power of wireless economics begins to make digital connectivity viable for the first time ever in parts of the developing world, those regions are also becoming accustomed to the first-order effects of digital connectivity. Beyond allowing distant emigrant workers to cheaply reconnect with family in their home villages, connectivity is democratizing access to critical information for people who have struggled without independent news of election results or fair market prices for their goods. As we'll learn more about later, it's also supporting the launch of banking alternatives for people who don't have to have access to a bank.

As much as these network-enabled activities may feel mature today to their beneficiaries, the core point of this book is that we are only beginning to embark on a global communications transformation. The changes in our lives and our businesses from the Anywhere Revolution are really just getting under way. As the network expands its reach and grows in capacity, the size and scope of the changes we experience as a result will actually accelerate.

The advent of television is a case in point. In 1949, there were just over a million televisions in the United States, or one TV for every 150 people. By 1959, that number had risen to 50 million. That brought the proportion of TVs to people in the United States to close to one in three. In the process, television fundamentally changed consumer and business behaviors far beyond the electronics industry itself, which of course was enjoying the popularity of its exciting new product. It created entire new business opportunities, such as television broadcast networks, and reinvented existing ones, such as advertising. Once there was one TV for every three people, U.S. television advertising rose to $1.5 billion in 1959, equivalent to almost 10 times that amount in 2010 dollars. And along with its commercial contributions came social influence; historians often credit television for swaying U.S. voters to elect John F. Kennedy to the presidency in 1960.

In the communications industry, this phenomenon has a name: Metcalfe's law. Dr. Robert (Bob) Metcalfe, one of the inventors of a seminal network technology, made the first attempt to quantify the tremendous rise in the value of a network from the addition of entities to it. He said that the rise in the value of a network is proportional to the square of the number of things it connects. Consider this: How useful is a phone with no one to call? The more people you can reach with a phone, the more likely you are to want one. This simple principle has been, and continues to be, the most fundamental driver of network expansion worldwide. Walter McCormick, of the United States Telecom Association, asserts a similar thought: "Network ubiquity is inevitable. That's because the utility of any network is in its ubiquity. If you go back in the United States to the original Communications Act of 1934, the idea was that a network that had very high penetration in urban cores had greater utility if it could connect people there with people in rural areas." Can we continue to expand the Anywhere Network beyond the people it touches today? "There are no real technological impediments. It would be contrary to our entire human experience to see a network of this magnitude not move toward universality," says McCormick.

If you accept that the expansion of the Anywhere Network is inevitable, if you're beginning to agree that we may face even more change ahead in communications from its rise, and if you're at least curious about the opportunities you may have to profit from those changes, then you will probably want to think about how close we are to Anywhere and how we will know when we get there.

How close are we?

When will we get there?

In fact, the answers to these two questions about Anywhere are critical for all businesses. Network operators need to invest time and money to deliver the capacity and experiences their users want. They need to build or upgrade networks, and to do so, they must be able to gauge when these very expensive investments will pay off. And businesses that will use the Anywhere Network need to know when and where it's appearing. The opportunities for profit are phenomenal, but everything from digital picture frames to location-based advertising depends on markets of connected consumers and businesses. Even governments considering broad-based digital services for their citizens face a similar chicken-and-egg problem: offering tax collection, permit issuance, and benefits registration online make sense only if all citizens have consistent, reliable access to a quality network.

As Nicholas Carr points out in his book *The Big Switch: Rewiring the World, from Edison to Google,* similar uncertainties arose during the build-out of electrical infrastructures in the early 1900s. Businesses didn't make investments in equipment that used centralized electricity until they were reasonably sure that there was reliable, affordable electricity available where they operated. Businesses didn't invest until they knew there was Anywhere electricity—ubiquitous connectivity to an electrical grid.

So where should you look for Anywhere? Where should you be selling and to whom in order to capitalize on the Anywhere Revolution?

Let's imagine that you're the CEO of a hypothetical start-up developing an after-market product for cars called GreenDrive. The device tracks a car's location, its fuel economy, and its engine performance data over the Anywhere Network and sends the information to the car owner and whoever is driving the car to help them make choices about fueling and maintenance. The product itself is ready to go, but before you introduce it to the market, you have to make two decisions: when to introduce the product and where to launch it. Because GreenDrive requires ubiquitous connectivity, introducing it too soon will leave customers without the information you've promised to gather for them, just as early buyers of televisions in 1946 found themselves staring at test patterns because of the paucity of programming. Introducing it too late means that competitors might establish a first-mover advantage.

As a start-up firm, or even as a brand-new division of a company on a short leash to see if it should win more of the parent company's resources, your team can afford to launch GreenDrive in only one country, because you plan to use the profits from that launch to fund your next launches in other countries. This means that you must also choose the country that has the best Anywhere Network infrastructure, since this is where the product is likely to be the most successful.

What you need—what any business needs—is a measure of how the Anywhere Network is advancing around the world. By assessing and predicting the progress of the Anywhere Network in each country around the world, we can answer these questions and more. This chapter introduces a simple but powerful measure of the pace of our connectivity revolution: the **Anywhere Index**. I'll show you how it answers these two questions for our imaginary product GreenDrive, and how it can help your own decision making about Anywhere products and services.

Measuring Anywhere

The Anywhere Index is the measure that Yankee Group analyst Carl Howe has developed, using the data we collect about available network connections in each country, to track the pace of Anywhere around the world and to predict when it will advance in significant ways. This tool now assists us in helping our clients plan market entry for connectivity endeavors.

It depends on a shared understanding of several key terms, so let's get that out of the way before I show you what it tells us.

First, it only looks at *digital* connectivity. This means that it counts network connections that can transmit IP traffic. Some older networks' infrastructure doesn't yet do that; for our measure we're not interested in those. Second, it specifically measures only *broadband* digital connections, which for our purposes today we have set at a threshold of at least 384 kps, which is the theoretical capacity of one particular wireless broadband technology still in use. Third, it measures both **wired** and **wireless** broadband connections. That's because wired and wireless connections are equally important to transforming a society with a pervasive network. Wireless networks can go where fixed-line networks can't, while fixed-line networks typically have more capacity and already exist today in many regions. Ultimately, as individuals, our desire for a complete Anywhere Experience will be best met by a combination of both wired and wireless connectivity wherever we are, but for now many of us will be dependent on just one or the other—so we count both. Fourth, it measures *potential* connectivity rather than actual use, since many wireless broadband connections are to mobile phones still used by their owners primarily for voice.

Finally, we compare the number of broadband fixed and wireless digital lines in a region to that region's population to get a comparative result: roughly how many people have reasonable access of some kind to the Anywhere Network. We express that ratio as a percentage.

Thus the Anywhere Index is the number of wired and wireless broadband lines per person in a country or region.

Using our latest data, the world's overall Anywhere Index in 2010 is 22 percent. This means that the world has about one broadband line for every five people. The reach and capacity of the world's networks has been growing steadily for the last decade—the Anywhere Index in 2004 was just 4 percent—but this clearly shows how far we still have to go in reaching all of the world's population with some form of broadband connectivity.

The Three Stages of Anywhere

Given our Anywhere Index definition, is Anywhere an all-or-nothing proposition? Not at all. The most significant changes we expect with the advent of an Anywhere society will come with universal connectivity—Metcalfe's law is the core reason. But when connectivity in a region rises, substantial value is created and important changes begin to take hold, as most of us have already experienced personally in multiple ways.

Yankee Group sees geographies moving through three stages of Anywhere, based on their populations' connectivity. (See Figure 3.1.)

1. **Emerging:** Emerging geographies are what we call those with less than one broadband line for every three people, or an Anywhere Index of less than 33 percent. Network connectivity is somewhat rare in these societies, and physical location is still vital to connecting with family, friends, and coworkers. Most information reaches people by print and broadcast media and therefore is dis-

	Emerging	Transforming	Anywhere
Connectivity	Rare	Frequent	Ubiquitous
Connected devices	One or two	Several	Many
Media	Physical	Mixed physical and digital	Digital
News	Scheduled	Mixed scheduled and on-demand	On-demand
Payments	Cash	Electronic/cash	Mobile payments
Services	Mass market	Segmented	Personal
Social interactions beyond immediate family and neighbors	Rare	Frequent	Constant

Source: Yankee Group

Figure 3.1 *The three stages of Anywhere.*

tributed on some sort of periodic schedule rather than being available on demand. For some people, access to a broadband connection at work or in an Internet café brings them resources that would otherwise be out of their reach, such as up-to-date news from distant or unofficial sources, or recent clinical test results concerning a family member's illness. But most large-scale systems and activities in emerging regions have yet to shift to being network-based. Fortunately, particularly in emerging regions with large populations such as India, forces including their own governments as well as commercial enterprises that view them as an important market are working to accelerate the pace of connectivity change in the region—and it's in these areas that we will see some of the most rapid expansion in years to come.

2. **Transforming:** Transforming geographies have at least one broadband connection for every three people in the region, giving them an Anywhere Index of 33 percent or more. Network connectivity at this stage is frequently available and is even the norm for many people and businesses, but it is not yet universal; usually it's the more rural or less affluent areas still struggling to get affordable broadband. But as in our example about the rise of television penetration, now that the crossover to one-third penetration has been passed, the nature and economics of connectivity are in substantial flux. In these societies, many people have more certainty about being able to reach friends, family, and work when they need to than people in emerging regions do; they're gradually becoming liberated from physical locations. But because the broadband environment is not yet ubiquitous, specific physical locations such as offices and meeting spots still feature into decisions about what to do and where to do it. Similarly, daily information is a mix of digital and traditional media, and scheduled and on-demand media. Behaviors are beginning to be transformed, and new services and experiences are starting to emerge. Some industries are in the process of making the changes they need in order to become Anywhere Enterprises, while others are still in the dark, thinking they might escape the impact. (See Figure 3.2.)

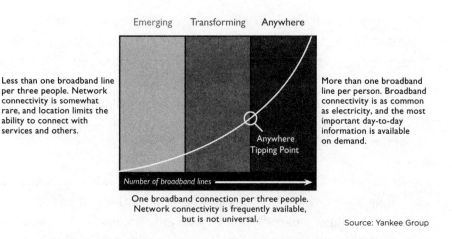

Emerging   Transforming   Anywhere

Less than one broadband line per three people. Network connectivity is somewhat rare, and location limits the ability to connect with services and others.

More than one broadband line per person. Broadband connectivity is as common as electricity, and the most important day-to-day information is available on demand.

Anywhere Tipping Point

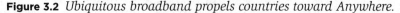

Number of broadband lines

One broadband connection per three people. Network connectivity is frequently available, but is not universal.

Source: Yankee Group

Figure 3.2 *Ubiquitous broadband propels countries toward Anywhere.*

3. **Anywhere:** Geographies that in Yankee Group's view are fully Anywhere are those with an Anywhere Index of 100 percent or more, meaning they have at least one broadband connection per person. This crossover to complete ubiquity relative to population is the **Anywhere Tipping Point**. In these regions, broadband connectivity has become as common and affordable as electricity—and just as unremarkable, because its presence and capabilities have become woven inextricably into people's lives. For the people living in an Anywhere region, it's very easy to be connected to people and devices. They expect to communicate with friends, family, and work at any time and from any place. The most important day-to-day information, everything from news reports to train schedules, is digital and available on demand. Virtually any activity they undertake as a consumer or a worker can be done anywhere they are on a variety of devices. Connectivity in an Anywhere region isn't static, though; the low cost and high quality of the network continues to give rise to more experiences that depend on the network and more devices that incorporate connectivity just as they use electricity. As a result of near universal connectivity, regions that have passed the Anywhere Tipping Point have already taken many dramatic steps away from the twentieth-century ways of interacting. Newspapers are fading away, and cash is less fre-

quently used. Previously stable sectors of the economy such as the entertainment industry are experiencing major shockwaves as consumers expect their shows and movies to appear on any device they like at any time. Businesses that market to consumers are more concerned about learning what they're doing at a particular moment than about obtaining more approximate representations of their interests such as age and gender.

Regions of the world that reach the Anywhere Tipping Point will ultimately do two things to our Anywhere Index. First, they will drive their index values well beyond 100 percent, because the number of devices in each person's world that connect to the network will grow as they add additional phones and netbooks, but also connected cars, clock radios, and more. How many broadband connections will the average citizen of an Anywhere Economy have? Considering the number of devices that will have connectivity added to them, which we discuss in the next chapter, the connections in your home, your car, your office, and on your person could easily surpass 10 or more.

Anywhere regions will also reset our definition of broadband, pushing activities and thus the demand for more network capacity even further. In 10 years, our Anywhere Index will have to be expanded to accommodate new phases of connectivity and new concepts of broadband.

Given this phased way of looking at the emergence of Anywhere, our global Anywhere Index of 22 percent would classify the world as emerging. But, as my scientist father was fond of pointing out, if you have one foot in a bucket of boiling water and the other in a bucket of ice, it doesn't mean that on average you're comfortable. Our global Anywhere Index doesn't tell us the full story, since in reality the state of access to the Anywhere Network varies tremendously by region. (See Figure 3.3.)

By region, North America, which lagged behind Western Europe in prior years because of its slower expansion of wireless broadband networks, now edges out Western Europe with an Anywhere Index of 87 percent. This is the result of accelerating expansion of both fiber and wireless broadband connections.

Adaptability is simply a region's pragmatism to adapt behavior, policy, and communications infrastructure to specific local conditions. In South Africa, disputes between current holders of radio bandwidth spectrum and the government are delaying progress in new spectrum allocations, and ineffective management of anticompetitive behaviors remains a significant barrier to progress. In Romania, on the other hand, a fast-growing competitive environment is enjoying additional wireless spectrum licenses that were recently awarded.

Collaboration means that entities in the region have the willingness to share resources with other market stakeholders. Even though both the United Kingdom and France are subject to the same general European Commission policy concerning telecommunications, they interpret and apply the policy differently. The United Kingdom has focused on restructuring the primary existing network provider, British Telecom, while the government in France has been championing infrastructure sharing as a centerpiece of its policy. Residents of France currently enjoy a much more comprehensive fixed broadband infrastructure.

Finally, regions with experimentation show the intuition and support to pursue experimental solutions in the course of expanding communications options. South Korea's leadership in broadband deployment and uptake is undisputed, due in part to the interventionist role of the government and its willingness to use public funding to drive progress.

Given all that, who's next? By 2015, 30 countries will exceed 100 percent on the Anywhere Index. (See Figure 3.4.) Some other highlights include:

- Nearly all of Western Europe will reach Anywhere status by 2012.

- Latin America will begin transforming in 2011, with Venezuela, Argentina, Peru, Chile, and Colombia each able to boast of an Anywhere Index of over 33 percent by 2015.

- Asia as a region will have moved into the transforming stage, with several new countries moving into the Anywhere stage: South Korea and Singapore.

And by 2020, we expect a significant further increase not only in the number of Anywhere countries, but also in the number of users

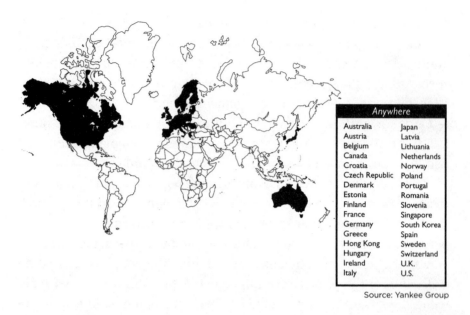

Anywhere	
Australia	Japan
Austria	Latvia
Belgium	Lithuania
Canada	Netherlands
Croatia	Norway
Czech Republic	Poland
Denmark	Portugal
Estonia	Romania
Finland	Slovenia
France	Singapore
Germany	South Korea
Greece	Spain
Hong Kong	Sweden
Hungary	Switzerland
Ireland	U.K.
Italy	U.S.

Source: Yankee Group

Figure 3.4 *Anywhere Index in 2015.*

able to access the network at speeds of 100 mbps, today's state of the art. Consumers in some select markets, most likely Japan and South Korea, will be able to access the Internet at 1 gbps or more. (See Figure 3.5.)

Back now to GreenDrive, our automotive connectivity opportunity, and using the Anywhere Index as a way to plan the expansion of connectivity products while the global network evolves. If the product had first launched in 2010 in South Korea, the choice for the next market might be Singapore. Certainly the state of a market's broadband network isn't the only factor in such a critical decision. We might have to look at regulations, the highway system, and the distribution of car types. But ensuring that the plan takes the state of the evolving Anywhere Network into account will be essential, as it would be for any profitable connected product.

By 2015, we expect the Anywhere Network Economy to swell to more than $900 billion in size. Investments in Anywhere will accelerate a move away from older products and services, but the network will bring greater prosperity to the world, particularly to those getting connected for the first time. Higher per capita incomes correlate with our Anywhere

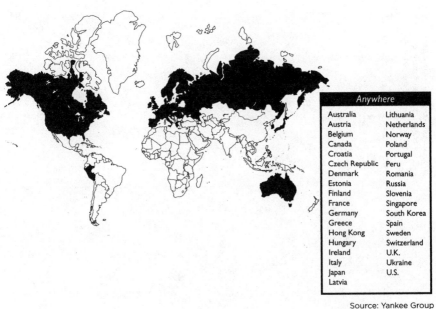

Anywhere	
Australia	Lithuania
Austria	Netherlands
Belgium	Norway
Canada	Poland
Croatia	Portugal
Czech Republic	Peru
Denmark	Romania
Estonia	Russia
Finland	Slovenia
France	Singapore
Germany	South Korea
Greece	Spain
Hong Kong	Sweden
Hungary	Switzerland
Ireland	U.K.
Italy	Ukraine
Japan	U.S.
Latvia	

Source: Yankee Group

Figure 3.5 *Anywhere Index, 2020.*

Index—that is, the higher a region's Anywhere Index, the higher that region's per capita income. Which is cause, and which is effect? The causality works both ways, as Bob Metcalfe points out: "The more prosperous you are, the more connectivity you can afford; and the more connected you are, the more prosperity you acquire. It's a chicken-and-egg question." It's also, thankfully, a virtuous circle. (See Figure 3.6.)

Connectivity can actually increase income per person. And anecdotal evidence, bolstered by recent research from the World Bank and the International Telecommunication Union (ITU), says that there may be even more causation here than we might think. Telecommunications has long been considered a facilitator of commerce, but the stories of mobile communications transforming even very small businesses are encouraging. Stick with me, and you'll see a few ways that this happens in the next section on consumers and connectivity.

A 2005 London Business School study claimed that a country's gross domestic product rises by 0.6 percent for every 10 mobile phones per 100 people. More than just greasing the wheels of the new economy, mobile communications services are expanding access into previously disen-

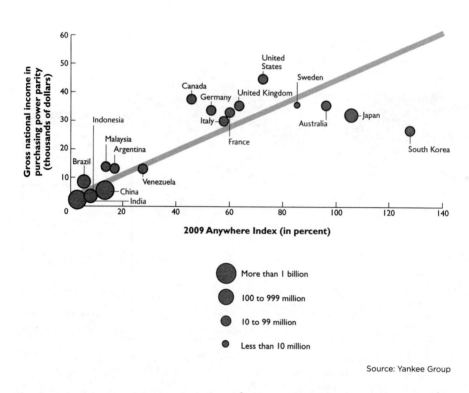

Source: Yankee Group

Figure 3.6 *Anywhere economies generate more wealth.*

franchised populations, which leads to growth. Safaricom's M-PESA (short for "mteja pesa," or "mobile money" in Swahili) banking transfers brought banking and payment services to Kenyans, 90 percent of whom don't currently have traditional bank accounts. Mobile phone penetration in Kenya is nearly triple that of bank account holders, though, and M-PESA brings these users the utility of wire transfers and micropayments for everyday activities such as transportation and food purchases.

We might chalk up these effects to basic mobile phone service and **simple message service (SMS)**, but this is just the first step up the technology ladder. The Anywhere Network of fixed and mobile broadband opens the richness of the Web and more complex e-commerce applications beyond large corporations and developed markets.

As countries begin to achieve Anywhere status, consumers will perform digital transactions more frequently. The Internet increased the pace of commerce by enabling one-click buying from any computer. Today, Internet transactions account for 3.4 percent of all retail sales in the United States, according to the Federal Reserve Bank of Boston. So guess what happens when you allow one-click buying from Anywhere—on mobile devices, gaming devices, automotive dashboards, and touch screens in public places? Consumers buy things more frequently. This will be especially true as consumers start buying groceries or paying for parking by pushing a button on their BlackBerry—already a reality in some places in the world. So much more convenient than having to fish for change.

Something else to expect, however, is that we'll all be more distracted. With the Anywhere Network providing ubiquitous connectivity, consumers will multitask like never before. Researcher Linda Stone talks about a new paradigm growing out of constant connectivity—"continuous partial attention," when we are continuously connected but paying only intermittent attention to multiple activities. Some 13- to 17-year-olds currently consume more than 24 total hours of media in the course of one day by multitasking. As the Anywhere Network reaches more of the world's population, multitasking and its associated continual partial attention won't just be a teenage fad. It will affect nearly everyone.

At the same time, enterprises will find the Anywhere Network transforming their businesses; as a result, the significance of corporate headquarters and their historic affiliations with particular cities may wane. Will BNP Paribas really be tied in any meaningful way to Paris, or Wrigley's Gum to Chicago? Anywhere means that businesses should be able to operate in any location. But as connectivity continues to vary wildly with geography (even as it expands), poor connectivity will be the death knell for start-ups trying to deliver broadband media and services. Don't be surprised to see the next generation of Anywhere start-ups boasting headquarters in Japan, Sweden, and Italy to tap into Anywhere populations.

The advance of Anywhere will drive critical market decisions. Despite the attraction of a high Anywhere Index in some geographic areas when it comes to launching next-generation connected products and services, there's even more money to be made helping what many call the "next three billion"—the people around the world who are not yet connected to the Anywhere Network.

Companies will target emerging or transforming economies based on the value of the consumer connection and the products they offer. Products that help consumers get on the network—think basic hand sets, cable modems, and routers—already find that their best growth opportunities now lie in emerging Anywhere countries. Those that add value to that connection—think digital music jukeboxes, TV streaming systems, and telemedicine—will naturally find a bigger audience with more profit potential in transforming Anywhere Economies.

Governments will find Anywhere Networks affecting their policies as they attempt to exercise more control. In a world where governments don't know where the next big threats will come from, the Anywhere Network's concentration of messages, transactions, and location information is just too tempting a conduit for governments to ignore.

Absent strong laws prohibiting it (or in the United States' case, in spite of such laws), government intelligence activities will focus more attention on the Anywhere Network. And as we've already seen in China where the government there often works to suppress access to sites that criticize it, some governments will also try to blunt its effects by restricting content available to citizens.

Many countries with a high Anywhere Index today got there by encouraging and providing incentives for widespread broadband deployments. Yet getting a consensus around these policies is anything but a simple process, and at the same time, network investments can be exceedingly costly too. All part of the time it will take for Anywhere to circle the globe.

* * *

In this first part of this book about the emergence of global connectivity, or Anywhere, I've set out a few key ideas for you to consider. Before we move on, here they are in a quick recap:

1. What I call Anywhere is a global transformation brought about by a common digital network, unrelenting and accelerating demand for broadband capacity, and powerful wireless economics.

2. Anywhere is a revolution in that the landscape around us will look very different as it unfolds, with shifts in power, behavior, and money all over the world.

3. What's emerging inexorably from this revolution will be a global digital network that is seamless, secure, pervasive, and intelligent. It will give rise to exponential new value creation in the form of new goods and services that depend on ubiquitous connectivity.

4. This Anywhere Network is not maturing at the same pace everywhere around the world, given vastly different starting points, along with differing levels by country of communications adaptability, collaboration, and experimentation. Knowing where it's developing is critical to anyone who hopes to participate in the transformations ahead.

As the Anywhere Network emerges, so do our opportunities to profit as businesspeople. We've now actually seen our first arena: building out the Anywhere Network. But beyond the building and powering of the network itself—the wires, gear, phones, and basic services—are profits for those who mobilize and capitalize on this global communications fabric as it unfolds.

The next two parts of this book delve into the two main areas of opportunity for Anywhere profit: the impact of this expansion of connectivity on consumers and on enterprises. In both cases, we'll look at how they're changing and then begin to expose your opportunities to deliver Anywhere products and services and to apply Anywhere technologies to reduce expenses.

PART II

The ANYWHERE Consumer

Chapter 4 | The Emerging Portability of Experience

"All devices will be connected; they'll have to, to perform the work we want them to."

—Glenn Lurie, president, Emerging Devices, AT&T

One of the simplest ways to define the Anywhere Revolution is this: the inexorable *diffusion* of digital connectivity. The global network is reaching further and further into our lives to touch new locations, people, and things every day. And when I say "things," I mean, ultimately, everything that matters to us, from tools in our homes, our cars, and items in the environment around us like lampposts and bus stops, to any important objects in our commercial world. The Anywhere Revolution is not only about PCs and phones connecting all of us to the rest of the world, but much more. I've claimed that this revolution is big. We just took a look at how and when the next several billion people will join the network; here we're going to understand how literally *trillions* of devices will too, causing the network's value to all of us to rise exponentially. In this chapter, we take a look at the diffusion of connectivity into more of the things in our lives. And as a businessperson, you should see the next commercial component of the Anywhere Revolution: new profit opportunities for your enterprise, whether you make things that people use, or help other companies do so.

Some Anywhere Things— Today and Tomorrow

The economics of scale guide many technology revolutions, and Anywhere is no different. The dramatic expansion of connectivity we have begun to experience in the past decade, and the further explosion still ahead, is fueled in part by the steadily falling costs of adding communications capability to devices. In 2001, Wi-Fi chips cost one developer of connected products over $120 each; seven years later the unit price for similar quantities had fallen to $7, as revenues from annual shipments of those chips climbed worldwide to over $350 million in total revenues in 2008, according to the Wi-Fi Alliance. That phenomenon will be a major driver of the connectivity expansion ahead.

In our tour of Anywhere devices, let's start with a quick look at computers. It's worth making sure that we see how these core Anywhere devices are insinuating themselves more deeply into our lives.

The first steps in portability in computing were made with the phenomenally successful introduction of laptop computers. In the last few years, a new kind of computer has been gaining traction in the mass market, the netbook. In December 2007, the first true netbook computer made its debut, the original Asus Eee PC 701. The PC industry had speculated off and on for years that we might want these smaller, less expensive devices, but it wasn't until the advent of Wi-Fi and mobile broadband, coupled with lower costs for computer components like memory, processors, and operating system software, that these types of devices could really take root. Netbooks are a ground-breaking confirmation of Anywhere's emergence, because these little computers are entirely predicated on connectivity. They are typically equipped with reduced functionality and much less onboard storage because they presume the availability of the entirety of the Internet to supply and store both applications and data. The result is a useful compact computing device that can sell for just a few hundred dollars—a breakthrough price point, lowered even further when a network operator like Verizon Wireless or Vodafone subsidizes the price to the consumer by bundling in a mobile broadband service contract.

When everything we do with computers can be done online, the need for multiple applications on our computers is eradicated or at least dramatically reduced. From about 10 million units in 2008, shipments of the broader category of ultra-mobile devices—including ultra-mobile PCs, netbooks, and **mobile Internet devices (MIDs)**—are expected to exceed 200 million units by 2013. Laren Whiddon, the vice president for wireless merchandise at electronics retailer Radio Shack, reports that the devices were a source of a huge increase in store traffic from the moment they were first introduced. As he explains, "We were the first retailer to bring out a subsidized model. Consumers are willing to forgo some of the laptop functionality to get the better mobility these [devices] offer." In some cases these smaller units are replacing aging laptops, but their very low cost is bringing Anywhere computing to many new users—students, lower-income families, and more.

But what about the broader picture of Anywhere devices? Connectivity has already begun to merge with many things in our world beyond the PC and phone: in our homes, in our cars, everywhere we go. Falling prices are part of a virtuous cycle in product adoption; another component of that cycle is ease of use. For many of us, first introduced to Wi-Fi on our laptop computers, it's now relatively easy for us to understand and use when it appears in other products.

In 2009 we began to see Anywhere's value broaden in the nature of connected devices available on the mass market. Connected cameras include the Panasonic Lumix with built-in Wi-Fi, letting users upload photos directly to Internet-based photo sites instead of connecting the camera to a PC first. Digital picture frames like the PhotoVu family simplify tremendously the process of getting those digital photos into the frame by allowing those same photo sites to be the source.

The Amazon Kindle and Sony eBook Reader have moved electronic readers from a questionable novelty into a viable, growing product category with the addition of built-in wireless connectivity with **prepaid** network access that the owner doesn't have to purchase separately. Steve Haber, the president of Sony Electronics' digital reader division, laughed when he acknowledged the market's early skepticism about the electronic book concept. "It wasn't too long ago that people were saying 'eBooks? What*ever*.' But when you add wireless to an open platform, it becomes an out-of-body experience. It's about immediacy, instant gratification." Amazon has enjoyed success with the Kindle too: In the spring of 2008, Citigroup analyst Mark Mahaney predicted that Amazon would generate between $400 million and $750 million in revenue from the Kindle in 2010, or between 1 and 3 percent of the firm's total revenues. But within just a few short months of his initial prediction, the analyst, noting the success of the device, had to revise his figures upward substantially, predicting that Kindle eBook sales were likely to reach $1.1 billion or 4 percent of Amazon's revenues. Connectivity changes the nature and opportunity to read. When a businessperson boarding a cross-country flight can choose from among 190,000 books instead of the 100 titles available at the airport bookstore, he views reading in a completely different way.

Similarly, connectivity is now transforming a more established electronics category—handheld game devices. Models like the Sony PSP Go are wirelessly connected, eliminating cables and bringing the user into a rich universe of multiplayer games. Gaming expert James Brightman calls the addition of connectivity to devices a "massive renaissance." Bluetooth technology provides short-range, fast wireless connectivity between multiple portable gaming devices, and Wi-Fi is making its way into handheld units to provide higher-bandwidth wireless connectivity over a larger area. Connectivity will make gaming an Anywhere experience because, as Brightman puts it, you'll be able to "have a quick throw-down against a friend in the car, or connect to an online gaming network through your home access point or favorite hot spot." Your gaming experience will be unaffected by location. Microsoft's experiences with the Xbox gaming family prove the point: in the past few years, as unit sales climbed at over 90 percent CAGR (cumulative annual growth rate), the proportion of buyers who joined Xbox LIVE, the connected online gaming environment for players, expanded even faster, at 130 percent CAGR. (See Figure 4.1.)

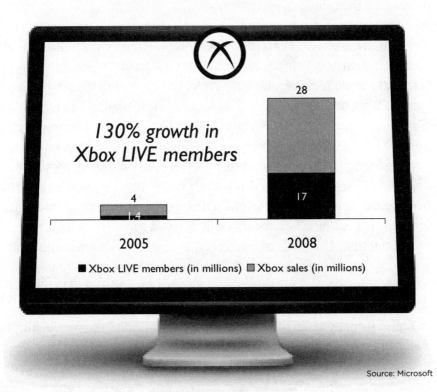

Source: Microsoft

Figure 4.1 *Xbox LIVE members.*

Another active Anywhere arena is the car. General Motors' OnStar system, currently with more than 2 million subscribers, is an in-vehicle safety and security system that gives U.S. drivers around-the-clock access to OnStar's trained advisors, connection to services for emergency assistance, and hands-free calling from within the vehicle. Satellite radio depends on digital connectivity. And the introduction of onboard navigation systems, which use satellite connectivity to determine the device's current position and to provide mapping updates, threaten to forever mute the husband-wife argument about asking for directions. But consumer-accessible connectivity in the car is just the tip of the iceberg. Since 2007 new cars in the U.S. market are required to have a digital readout that tells you when your tire pressure is low. How does the computer system in your car know when the tires are low? You can't have wires connected to spinning wheels; instead there are chips embedded in the wheels that transmit a wireless signal to a receiver in the onboard computer.

New Anywhere devices that reach the shelves of stores around the world in the next two years will largely be the result of product development already well under way now. Sony's Haber expects electronic readers to explode as a category. "They'll come in different shapes and sizes. Today, it's a glass-based product, not designed to be thrown across the room. As development expands, it can become more durable, more flexible—maybe even bendable. For younger users, you'll need immunity to juice boxes." AT&T's Glenn Lurie sees a lot of near-market-ready connected products in his role as president of the firm's emerging devices division, which partners with Sony to supply the bundled connectivity for its Daily Reader. "In 2010, we'll see a lot of competition. Everyone and his brother will have a lightweight computing device, whether you call it a netbook or whatever. The next area after that will probably be portable navigation devices. People thought these would slow down in sales. But when you can deliver a lot of data to them, it really changes what they are and what they do. Real-time traffic information, points of interest . . . they will be bigger than ever."

Some of the devices we'll be seeing include connected televisions, bringing our home broadband connection directly to the display. Hilmi Ozguc, a successful pioneer of broadband video services at Maven Networks (now a part of Yahoo!), expects big things. "People are already beginning to patch together broadband TV viewing experiences using their PCs and Web sites like Hulu that are as good, maybe better, than the conventional TV viewing experience." Yankee Group's own consumer surveys in the United States pointed to less than 5 percent of the population connecting their TVs to their PCs in order to view that broadband content on the larger screen in 2009, but the change is afoot. The next step will be for wireless connections from your TV to the network; ultimately the need for a physical cable to make the connection will fade. Ozguc's view is that the Internet in many ways will be a savior for TV. "The industry is suffering from the fragmentation of its audience in the living room during prime time. The building blocks for a better TV are there; we're just waiting for the match to be lit." (See Figure 4.2.)

Will there ultimately be any device in our lives that won't be better off with connectivity? Once you start looking, the opportunities expand. "Weight scales, blood pressure cuffs—all of these home tools are naturals for embedded connectivity," says David Rose, a serial tech-

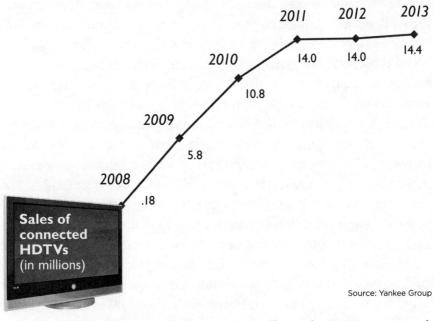

Source: Yankee Group

Figure 4.2 *TVs will soon be Internet connected.*

nology entrepreneur now leading Vitality, the connected pill bottle company that we discuss in Chapter 1. "They cost enough, in the $50 to $250 range, to bear the cost of an additional $2 for a wireless chip set. Being able to easily share data about your health with a care provider or a social network can be very helpful." Musical instruments? With the adoption in the industry of standards for encoding musical notation (versus recordings), connected instruments can receive music for their users to perform. Yamaha's Disklavier Mark IV piano has a Wi-Fi connection that lets the pianist download music for the piano to play, either automatically or as an accompaniment for the pianist's own performance. Garage door openers? Most already incorporate a wireless transceiver that communicates with a small remote control device in your car; why shouldn't the garage unit be able to detect the arrival of an authorized vehicle and open the door for you automatically? It can—and eventually it probably will.

If you want to create and sell one of the hundreds of Anywhere devices that we will buy over the next five years, a good way to start is by looking at the devices in our lives today and thinking about how connectivity can make them more useful to us.

We've been talking here about the development of Anywhere things that we'll acquire and use as consumers, making our activities increasingly portable. And as businesspeople, whether we're the sellers of those items or not, we'll want consumers using these devices because of the tremendous value to us of these additional digital pathways. It's worth mentioning before we go too much further that a less obvious but ultimately much larger arena of advancing connectivity is devices that we might not own ourselves but that we come into contact with all the time. Cash registers, vending machines, juke boxes, parking meters, casino slot machines, billboards, buses and trains, and many other things we touch every day are being augmented with inexpensive, powerful, wireless connectivity to make the experience we have with them faster, smarter, or more entertaining. And beyond what we touch as consumers, Anywhere is reaching things in industry, agriculture, transportation, and many other commercial settings—sending temperature readings of refrigerated medications in hospitals to pharmacists; tracking moisture levels in the ground to alert irrigation systems, or transmitting video imagery of activity throughout a busy port to the harbormaster's security team. There are even more profit opportunities in those developments than in the consumer arena, and we talk more in the next section about how Anywhere is changing the environment we work in.

For now, let's return to figuring out where the profit lies in successful Anywhere devices for consumers. We'll look at the components that will make Anywhere magic happen and then highlight the benefits to businesses that get it right, either in developing these items or in partnering with the firms that do.

Getting the Anywhere Recipe Right

Connectivity is being added to devices and experiences in our lives in much the same way that electricity came into people's lives a hundred years ago. Gradually more and more common tools were equipped with motors and cords that could be plugged into a standardized and affordable electrical grid to save labor. In the early twentieth century, consumers were offered a growing array of whiz-bang new electrical products that were essentially a motorized or electrified version of an already common household tool. In the kitchen alone, homeowners began to surround themselves with electric coffee percolators and can openers, electric mixers and blenders. Food preparation began to assume the availability of those devices, and today my own collection of cookbooks expects that I have a food processor.

But not all electrical gizmos enjoy that indispensability—the presumption that you can't manage without them. I own an electric knife, which I think I got as a wedding present, but to be honest, I couldn't tell you where it is right now. My best guess is that it's buried somewhere in the dark recesses of a corner cupboard in the kitchen. The reason? Very limited utility. It just doesn't do enough in a unique way for me to remember to haul it out in the few circumstances where it does a significantly better job than a regular knife.

You may carve more turkeys than I do, making the electric knife more useful to you than it is to me. Nonetheless, some Anywhere things will enjoy widespread adoption, while some will feel contrived to some of us. As entrepreneur Vanu Bose says, "Lots of things will be tried; some will turn out to be useful." When only a modest improvement over the things we already use is offered, combined with the need to replace current things, some Anywhere devices will have a very hard swim **upstream**. "My favorite negative example is the 'spork'—that weird all-in-one utensil supposed to do the job of three pieces of cutlery," says Vitality's David Rose. "I've pitched products to investors and retailers; I can just hear someone explaining how great it was going to be. 'See? Here's a rough edge where it's like a knife, here's a forkish sort of tip, and this curvy part is like a spoon.' Only it didn't take off. There is very little cost to having the single-purpose flatware we already own, and a lot of convention already associated with its use."

What types of devices and experiences lie ahead for consumers around the world? Which are most likely to fit with our appetite for portability and enjoy broad adoption? If you want to profit from the emergence of these devices, you'll need to give some careful thought to their creation or to partnering with the firms that do. At Yankee Group, our connectivity experts see three main dimensions to Anywhere device success: *benefit*, *use*, and *price*. To show how these dimensions apply, and how they fit together, we take a look at each one in turn and then identify where the Anywhere profit comes into play.

Building Up the Benefits

To avoid the fate of the spork relegated to Boy Scout camping trips and the electric knife only brought out for holiday birds, Anywhere devices must deliver a clear *benefit*. The near-infinite resources of the Internet mean that the benefit of Anywhereness coming to a device could be quite broad. Think about how you might staple that breadth onto your consumer focus, whatever it is. What additional benefits would your offerings provide that respect the nature of the offering itself?

Let's start with the essence of the product itself. Could connectivity make its core function more valuable? Tom Sebok is the president and CEO of Young & Rubicam North America, a leading agency for global brands. I asked for his perspective on connected products like Ambient's umbrella, which uses the wireless network to check the weather forecast, turning on a light in the handle when it may rain where its user is. He was amused at first and then thoughtful. "The lens through which I look at brands adding connectivity to devices is simply this: is it enhancing the essential experience the brand is meant to deliver? An umbrella that helps me know whether it's going to rain is more 'umbrella-ish': delivering more of what I want from an umbrella. It's enhancing its protection by increasing the chance I'll take it with me when I'll need it, while reducing a key pain point, identifiable in any brand relationship, by letting me know when I don't need to take it. That's where your focus should be in thinking about adding connectivity: Does it enhance a core benefit or lessen a current pain point?"

Let's move on to location. The connected umbrella in fact incorporates location in how it augments its umbrellaness. In fact, the essence of the Anywhere Revolution is to redefine the importance of location. Letting us integrate location information into the things we do, automatically, is possibly the most transformational aspect and immediate customer benefit of an Anywhere device. For consumer products, connectivity eliminates the barrier of distance between us and our desires and capitalizes on the advantages of proximity. Search, commerce, social networking, even rain protection is made more useful when the activity can incorporate knowledge of our whereabouts. Proximity to our destination, to the hard-to-find spice ingredient our recipe calls for, to friends we might be able to invite to join us for dinner—a world of **location-based services (LBS)** awaits.

Enterprises that seize on the ability to integrate location into our existing behaviors stand to reap the commercial rewards, while those that delay not only don't benefit, but could fade from our experiences all too quickly. In the hospitality industry, Zagat was once the gold standard for restaurant reviews. Restaurants could flourish or fail based on its ratings. Out of nowhere, a Web property named Yelp, with no particular expertise in restaurant reviews but a leader in mobile Web, has become a leading resource for foodies. Randall Stross of the *New York Times* leaves us in no doubt: "Yelp has crushed Zagat.com in the restaurant ratings game, with a fanatical adoption of **Web 2.0** social media techniques and clever marketing." A comparison of the two sites' restaurant-related traffic in Figure 4.3 proves the point. The rep-

Restaurant sites: unique visits
April 2009

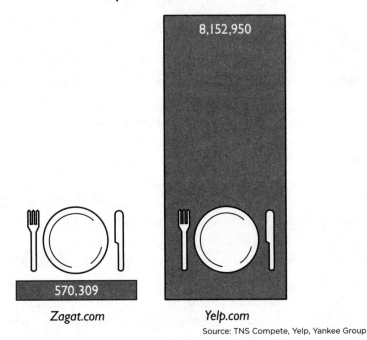

Source: TNS Compete, Yelp, Yankee Group

Figure 4.3 *The high price of missing the Anywhere phenomenon.*

utation of the Zagat brand doesn't overcome the increased relevance of Yelp's mobile service when diners are on the move without their paper guide but always with their phone.

More benefits for Anywhere devices lie in the ability for a connected product to gain updates or upgrades. Microsoft executives said in 2009 that they expect the useful life of the Xbox to be ten years. Most consumer electronics devices have an average useful life of two to three years. But because the new Xbox models are Internet-connected, they can be constantly updated with new software and features.

"Threading" is my own term for how connected devices can create comprehensive connected experiences. Larry Weber, longtime technology insider and chairman and CEO of W2 Group, talks about an evening following the death of author John Updike in 2009. "I spent an hour and a half on the Amazon site," says Weber. "They had some rare video of Updike reading his work. I watched that, then I read some reviews. I posted one of my own. I bought some books of his that I didn't have. What I had was a rich 'John Updike moment'—educational, social, commercial, everything in one." Not only do we appreciate being able to explore topics through multiple simultaneous activities, but we also appreciate it when the things we use help us maintain continuity across time—threading discrete events together to help us overcome the increasing interruptions in our attention. Readers of Kindle eBooks who own both a Kindle reader and an iPhone can stop reading something on their Kindles and resume reading it on their iPhones—on the same page where they left off.

Connected devices can know about the other connected things in your life, so they can cooperate with each other to thread together your activities into a more seamless series of episodes. Ultimately, if the appliances in my kitchen know I'm making a recipe from the Web site Epicurious, a connected microwave might know how long to defrost the frozen cherries, a connected mixer might be able to adjust its speed and timing as I add the cherries to the cake batter, and a connected oven could automatically start warming up while I pour the batter into the pan. They'd still be performing the essence of their value to me—only better. If you want to help drive the revolution coming our way in connected devices, your organization should ask this question: How could the things you offer to a consumer participate in the larger context of their use by threading themselves together with other devices, other components of a consumer activity, to provide continuity and coherence?

Another core aspect of the Anywhere Revolution is immediacy; having a pervasive network fabric lets us seize the moment wherever we happen to be. David Rose translates that into commercial opportunity by thinking about all the decision-making moments that a consumer has in the course of a day. "If you take the intersection of the important decisions you make and where you are at those moments," he says, "then I think an interesting question for connected devices becomes this: What's the available 'real estate' around you at that moment, and what could you know that would help you decide?" He pictures being at the kitchen counter when needing to decide if it's time to leave for the bus; upcoming bus information could appear on a surface in the kitchen. (If you think that's far-fetched, there's more than one municipality making its bus scheduling and real-time arrival data available to developers to build specialized applications to use it; the San Francisco BART program is one.) He goes on: "If my calendar knew I needed to leave for the airport based on current traffic and how I liked to go there, it could just say, 'David—car service, right?' The way I think about it is this: When would the perfect butler subtly appear and ask the right question?"

Anywhere devices can be green devices, too. Mike Muller is the chief technology officer for ARM, a pioneering semiconductor firm whose processors are in devices ranging from BlackBerrys and cable set-top boxes to cameras and flash storage drives, and more. He predicts, "There will always be stuff you won't do with your device because you don't care enough to bother." For instance, many consumers will still avoid paying attention to shifting rates for electricity when deciding when to run their dishwashers. "But when you don't *have* to bother—when the device you plug in has enough intelligence to talk to the grid and to other devices and figure out when it should do what it has to—then that's a breakthrough."

Identifying the Right Uses

The rich potential of Anywhere benefits could come through adding location, experiential continuity, immediacy, updated product capabilities, or much more. No matter what an Anywhere thing does, though, a clear worthwhile benefit comes to its user only through *use*. The world of consumer products is littered with those that are hard to use, so much so that core parts of their value aren't fully exploited. It's a wonder that VCR sales leapt from $389 million in 1979 to over $5.2 billion in 1986, while it became a well-established joke that most of their clocks blinked "12:00" constantly, clearly not being used to record timed programming from the TV. At least those devices sold anyway, because the core use—shove a tape in and press "play"— was straightforward. Adding Anywhere to a device presents a veritable minefield of blinking-clock problems. A mobile network operator recently piloting a home base station for wireless phones found that its users were unable to successfully enter a 24-digit password into a tiny onboard display when, for security reasons, the digits they entered disappeared immediately afterwards. For strategic decision makers, therefore, thoughtfulness about the way the consumer wants to use a device is essential.

Possibly the best use goal for any product, including Anywhere devices, is intuitiveness. We all have a frustration budget for learning new things. How much time we'll put into that depends on our personalities and the perceived benefit. A more thoughtful VCR design would have let those of us who have the time and energy to learn how to program a VCR, do so without advertising the incompetence of the rest of us through that steadily blinking clock. The truth is that our patience tends to be pretty low for consumer products, and products that appear overly complex have a tougher row to hoe in winning adoption. Steve Tomlin, a founder of Chumby Industries, creators of the Chumby platform for connected consumer devices, pointed at a bad example: "Kodak came out with a digital picture frame with a bunch of buttons around the edges of the frame. You had to mentally map points on the front screen to those buttons on the side to figure out what they each did. Nobody got that at all."

One secret to conceiving how an Anywhere product will work intuitively for the user is making an explicit break with the mental model of the computer and the Internet browser. The much-maligned "Internet refrigerator," launched in 2002 by LG, failed in part because it was a bit ahead of its time. But the core flaw was the assumption by the manufacturer that the general model of using the Internet that applies to the computer should also be the interface model for a kitchen appliance whose job is to cool food. A Web browser on the door? A more broadly successful connected refrigerator is surely in our future, but it will succeed because it places its connectivity explicitly in the service of the refrigerator's main job—helping us manage our perishable food. Would I like it to help me place a Peapod order? Sure. Would I like it to check the latest food safety guidance about how long I can keep that fish? Absolutely. But the way in which it performs those services needs to make sense when I'm at the fridge in store or retrieve food mode. Make the connectivity help the refrigerator preserve my food, and I'll want it.

The introduction of Apple's first iPod device in 2002 transferred that company's remarkable success in designing easy-to-use computers to consumer appliances. It wasn't the first portable music player that used the MP3 format to store and play music; an earlier product from Creative Labs provided me in 2000 with my own "aha moment" about the future of digital music—notwithstanding a confusing menu and a tiny screen with even tinier buttons. But the wave of ongoing innovation in the Apple iPod product line in the ensuing years has raised the bar for makers of all kinds of consumer devices. From the simplicity of the physical interface to the color choices and the packaging itself, consumers drool at the combination of brilliant simplicity and pure sex appeal. It's a lesson anyone wanting to add connectivity to a device ignores at her peril.

Fortunately, Mike Muller at ARM points out that connectivity can actually help *simplify* a device's use by reducing the need for buttons and tiny screens onboard. "I call it the split **user interface**," he says. "My high-end washing machine may have its own glass display built into it. But my hot-water heater, down in the basement, doesn't need that, and I'd rather just control it from my phone or my computer." Moving functionality away from the device to a mobile phone as a sort of universal remote control may become commonplace. Where could your firm's product designs save money and reduce use complexity by borrowing this concept?

Another critical feature of Anywhere use may be unintrusiveness. I visited an office recently where a digital picture frame sat on a desk, turned off. I mentioned how much I enjoy the one I have, and the embarrassed executive said, "I love the photos of my kids, but it's just too distracting." Computer screens are bright, cause eyestrain, and seem to demand constant attention even when they're in the background. In describing the thought process around the design of his company's first connected device, Chumby's Tomlin said, "I asked myself, could you take a large set of resources and make them personalized and available at a glance? The way you quickly look at a wall clock without interrupting your daily routine. Computers, even handhelds like the iPhone, require a more explicit disengagement from what you're doing. I feel many times I'm disrupting my life for the

Internet. I think connected devices are about bringing together your Internet life and your real life."

David Rose calls this quality "glanceability." "I want devices that adjust their resolution to my interest." He worked with the electronics firm Belkin and the French network provider Orange to create Live-Screen—a small, in-home wireless display with a proximity sensor. At a distance, its information display is very simple, presenting for instance just the current temperature, and is bright enough to read easily. As you approach, perhaps because you want to know more, the information it displays can increase in detail, showing the weather forecast for the rest of the day, while the brightness of the screen can reduce to be less glaring. How can you create experiences with products that incorporate connectivity but that allow users to moderate the degree of involvement they have with it? If you can, your overall exposure to the user will increase—the device will stay connected and in use, unlike that nameless executive's photo frame.

Pricing It Right

So we've identified the right benefits for our Anywhere device and planned an intuitive, unintrusive experience for its user. Anywhere devices that win will bring a clear benefit and appropriate use together with the right *price*. It's that same combination that has propelled TVs, computers, CD players, and virtually every other consumer product success story of the twentieth century. The unique challenge in pricing Anywhere devices is the added element of connectivity. While the connectivity technology in the device itself continues to become substantially less costly, we still have to account for the cost of the network access that the device needs. Someone, or some thing, has to compensate the providers of the Anywhere Network for the traffic that flows over it.

Successful Anywhere devices will solve the consumer pricing problem with creativity. Amazon had the insight to team up with Sprint/ Nextel for the simple bundled connectivity experience that has helped the Kindle get off the ground. The Kindle buyer is essentially prepaying for the cost of his network use in the up-front price of the device itself; Amazon then pays Sprint. As Sprint/Nextel CEO Dan Hesse sees it, "You could see Sprint partnering with a financial services firm to provide customer connectivity, doing their banking via mobile. Or a consumer electronics manufacturer providing a partial subsidy of the connectivity cost to deliver periodic functionality updates or customer service."

Yankee Group and most of the executives interviewed for this book agree: Consumers don't care to add new subscriptions to their lives. Very few new Anywhere products will succeed that depend on consumers signing them up with a network operator as if they were adding more mobile phones to their family plans. Think back to the explosion of electrical appliances in our homes. If we'd had to call the utility company to turn on a new device in a separate subscription, resulting in a bill that looked anything like my cell phone bill does these days, would we really have bought all those beard-trimmers and can openers? Probably not. Anywhere connectivity costs for the most part will need to be bundled into the device's retail price, subsidized by an enterprise that benefits from the ongoing connection, or, for less frequently used connections, charged directly to the user on a per-use basis. Yet another dimension in which Anywhere is revolutionizing how we do business. Is your firm ready to partner with a network provider to deliver a constant connection to your device's users?

Where Is the Profit in Anywhere Experiences?

Now that we've seen how benefit, use, and pricing will need to align to create successful Anywhere devices, let's look at how the rise of diverse connected devices will change how businesses make money. The profitability of consumer connections will change in three ways.

The first is in how enterprises *reach the customer*. If your firm has a service that consumers use, rather than a device they buy, you should be asking yourself which connected devices will provide critical pathways to reach consumers with your message or services. As TV audiences shrink (while the price that advertisers pay to reach them actually goes up, not down) and as consumers' comfort with ad-supported media experiences from sports to news is extended to new devices, many of those devices will present new channels for marketing and promotion. Their cost per prospect may be less than the media you use today. Given the ongoing nature of the connection to the user, your ability to know a lot about whom you're reaching and where they are can make your spending much more efficient—another aid to the bottom line.

The most important way in which consumer business profits will change, though, is simply this: how you *keep the customer*. If you're the maker of a connected device, you have the potential for a rich, long-term dialogue with its user. Marketing textbooks are full of calculations of the lifetime value of a customer because data across every industry show that the longer you hold onto one, the more money you can make, more effectively amortizing the expense you had in acquiring them. Do you make a consumer device today that you ship with a flimsy paper warranty card? If so, you're hoping against hope that buyers will take the time to fill it out and return it to you so you can understand just a tiny bit about who they are and what they want. "Keeping the customer" in the case of an Anywhere device means gaining the chance to know who customers are, to understand every day how they use the product, and to easily ask what they want from you next. And when the time does come for the user to replace the item, you might not lose the sale to another brand. "Keeping the customer" means you can refine your understanding of who likes what you do so you can focus more effectively on finding more customers just like that one.

Once you win and keep the customer and reap the efficiencies in your business from doing that—refining marketing and lowering your customer acquisition costs in the process, which helps the bottom line too—we come to the third way in which Anywhere is revolutionizing how you profit in a consumer business: how you *up-sell the customer*. Not only do you have a built-in path to learn who customers are and what they do, but you can augment what they get from you. Online retailers like Lands' End remember what we've bought or looked at and suggest related items—a handbag to match the shoes, for instance. Anywhere consumers who buy a connected camera might be interested in a software upgrade that enhances its digital zoom, postcard formats, or onboard cropping capabilities. But to capture the impulse purchase, that upgrade must be delivered directly to the device, without a trip to a store or a cumbersome transfer from a PC to the device.

The connectivity service provided by your Anywhere devices might come in two flavors: a basic service with its cost either bundled into the device price or subsidized by the provider of a basic service like a bank, and premium, where the device user pays directly for additional connectivity features. Subsidization in connectivity works in attracting consumers: Carrefour, the international retailing giant, works jointly with network operator Effortel to offer mobile services in stores in four countries. Shoppers in stores in Poland can use their Carrefour loyalty points earned through in-store purchases to get free mobile minutes. Why does Carrefour do it? "Everything retailers do, they do for margin," says Liudvikas Andriulis, Effortel's chief marketing officer. "Phone services have a much better profit margin than milk, meat, or tomatoes. Retailers don't have to build more stores, or find more shelf space, to make money by adding mobile services."

The Connectivity Diffusion

Ultimately, makers of virtually any consumer product will have to ask themselves: why *wouldn't* we add connectivity to this item?

If you can enhance the user's experience with sending, or getting, real-time information, you should.

If you can add value to the product with connectivity—perhaps contributing to the cost too, and thus defraying the price of the product for the consumer—you should.

If you can extend the life of the product in the consumer's hands by providing service or updating it with new features, you should.

And if you can partner with a firm that can do any of these things to bring your consumer service or message to more "surfaces" in the consumer's life, you should.

At the outset of this chapter, I claim that we'll be connecting trillions of devices to the global network. Now that we've looked at the variety of things in our lives that will gain connectivity, and the commercial value to enterprises from that development, I'd like to revisit that claim by doing a simple bit of cocktail-napkin math. It's not meant to be a forecast but rather a way of getting our thinking into the right order of magnitude.

Think about 2020. Estimate the number of personal items under the control of a typical person in the developed world where we'll have ubiquitous connectivity—say 10. It's not hard to get there when you consider a computer, a handheld device, a car or motorbike, a primary TV screen, and a few other displays or entertainment devices. To this, then add the items we're exposed to in a given day that don't belong to us but that will have connectivity by that time: traffic lights and public transportation; industrial assets; goods in stores and warehouses; equipment in offices, farms, and factories. On a per-capita basis, this could easily be another 100 items. Various estimates put the number of connected people on the planet by 2020 at around 6 billion. Ten personal items times 100 ambient items times 6 billion—I have a hard time keeping track of all the zeros, but I'm pretty sure that's 6 trillion. Our world is becoming pervasively connected.

This development will change us profoundly, combining our fundamental mobile urges with truly portable experiences. Chapter 5 will give you a preview of the Anywhere Consumers of the future. We'll take a trip around the world, combining our expectations about when and how the network expands with who and what will use it to profile the different ways we'll experience this change. In Chapter 6, we return to the present day to look at how these appetites are already being exposed, which means that you have to begin now to adapt your products and services for Anywhere.

Chapter 5

Meet the ANYWHERE Consumer of the Future

"We all just want to be a part of the world."

—José María Álvarez-Pallete López,
CEO, Telefonica International

Technology is a major catalyst of changes to the way we live and work. It has been so since the wheel came into our lives. And when it connects with fundamental human needs in cost-effective ways, its impact can skyrocket.

In Chapter 4, we explore how digital connectivity is diffusing into more devices and experiences. This accelerating phenomenon is bringing broad changes to both what we do and how we do it, no matter what part of the world we live in. Some of these changes are recent or already familiar to us, while others have just begun to gather steam and will mature over the next 10 years. But once connectivity comes into our lives, we will never be the same.

To get a sense of the breadth of this—the variety of activities and attitudes that the Anywhere Revolution will affect around the world—it's useful to take a quick trip to the future, previewing our connected lives once many of the changes we can anticipate have become commonplace. So this chapter takes a look at the connected lives of four different people 10 years from now.

At Yankee Group, we devote a lot of time to researching the pace at which connectivity technology will reach markets around the world. Our insights into how component technologies are maturing, how their costs are falling, and how receptive markets are, let us anticipate when those innovations will reach the market and counsel our clients about what role they can play in those developments.

To bring those insights to life, we take our assumptions about the emergence of the Anywhere Network and explore a day in the life of four imagined inhabitants of the future.

Why four? There are two reasons. First, we have to look at Anywhere Experiences across a variety of regions in the world, since the Anywhere Network won't look the same everywhere we go. In urban India, burying broadband cable in the ground where it's hardly available today will probably still be economically and logistically infeasible, but the population's enthusiasm for mobile experiences will have propelled strong expansion of the wireless infrastructure, particularly in highly dense urban areas. In rural Africa, wireless connectivity will gradually spread, with a rise in availability of solar-powered, diesel-powered, or other equipment that doesn't place heavy power demands on a limited electrical infrastructure. In 2009, Kenya's demands for better connectivity to the rest of the world were met in part by the arrival of SEACOM, an undersea cable delivering multimegabit connectivity to Mombasa. The economics of network expansion in both India and East Africa will be constrained by people's limited ability to pay for connectivity and connected devices. This makes it more likely that major brands will subsidize connectivity devices and services to drive loyalty as consumer purchasing power begins to expand. (See Figure 5.1.)

But in North America, today's comparatively rich fabric of both wired and wireless networks will evolve further, adding more bandwidth and smarts, which will engender much more complex, media-rich experiences for many users. Consumers may ultimately choose from several models to pay for their connectivity services: from free, to basic, to premium prices for premium service. And, in some parts of the world such as Japan, Korea, and in our last example in this chapter, Sweden, an extraordinarily evolved fiber-optic network will continue to be the platform for the world's most advanced connected experiences. Governments in some of those markets may regard the provision of high-quality networks as a public good whose access, features, and price need to be encouraged or tightly regulated. These variables will have a big impact on the nature of the Anywhere Network in those areas and what experiences it will support. But where the fiber-based network continues to evolve alongside the wireless

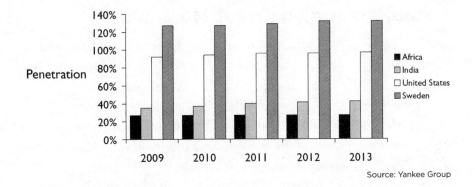

Source: Yankee Group

Figure 5.1 *Mobile phone penetration forecast.*

infrastructure, both growing in capacity and intelligence, the richest and most thoughtful Anywhere Experiences will emerge. They will allow consumers to move seamlessly between multiple network connections and devices, without regard for the details of the physical connections they use.

And if we want to have a really useful look at the future, we also need to take into account the incredible diversity of human beings. A core element of the Anywhere Revolution—the mobile phone—already touches more of the world's population than any other electronic revolution to date. As a result, it's already very clear that while we all share basic desires like connecting with each other, there are differences in what people do and think about mobility that reflect diversity in their education, income, lifestyle, age, profession, and more. Our prototypical consumers of the future need to range from teens to seniors, from farmers to merchants to office workers, and from those who are relatively well-off to those who are borderline poverty level. While these Anywhere Consumers will all be connected in many ways, the things they do and the impact that connectivity has on their lives will be different. And the ways your firm can profit will vary accordingly.

So let's visit with Habib, Fatuma, Sheryl, and Arne. As they share their stories, think about the commercial opportunities presented by what they do. Where might your company be able to contribute to their lives? I'll point out some themes at the end of the chapter; see if you spotted the same ones.

Habib, the Indian Street Merchant

On the busy streets of Mumbai in 2020, we find Habib, a sharp, charismatic teenager who recently finished school and became a street merchant like his father and grandfather before him.

He is working alongside his father for now, since he is only 15, but he has many ideas for the business. Habib wants to apply all sorts of new ideas to improve on the old business model. If only his dad were a little less old-fashioned.

Habib lives with his parents and his four siblings right in the heart of Mumbai. They're not wealthy, but they have a roof over their heads and just enough food on the table to feed everyone. In a country like India, having even this much makes all the difference.

Still, Habib has high hopes for his and his family's future. When he was at school, he learned a lot about the world. The school had wireless Internet access for students both in the computer labs and in the library. Students could get online whenever they wanted to during lunch break; Habib and his friends spent a lot of time chatting with other kids their age around the world via Webcam. Habib and his siblings were given netbooks to take home. They played games and did their school work on their computers; they even taught their father to read.

Habib learned a lot about the value of connectivity by interacting with his friends. It makes life easier, and that's really why he wants to incorporate it in his work. He'd like to see the business run on autopilot some of the time, because his father's getting older and it just makes sense.

Unfortunately, Habib's father has other ideas. The first battle between them was about what to sell. Traditionally, the family had been selling samosas (triangular-shaped fried pastries filled with spiced meat or vegetables) in Colaba, where there are plenty of tourists. This meant that Habib's father had to get up early every morning to go to the market, pick out groceries, and then pedal his cart in the other direction to his usual spot. Some days, the vendors didn't have what he wanted. Other days he knew he was paying too high a price, but he didn't want to go farther to look for a lower price because another pushcart would be in his spot by the time he arrived.

Habib wants a different life, not just for himself but for his whole family. In the morning, he opens up a netbook that the store chain Subhishka gave him for the points he earned from repeat food orders. He updates their inventory online based on what was used up since last time, and places orders for whatever goods he and his father need. When he pedals past the nearest Subhishka shop on his way to their favorite spot, his order is waiting for him. His account is directly linked to the store, so the debits are processed automatically whenever he orders food.

Habib picks up free Wi-Fi at his cart from a nearby hotel. To improve foot traffic, he offers mobile coupons; anytime someone walks within 100 feet of his cart, that person gets an advertisement on his or her handset for a free drink with the purchase of two samosas.

―――

Mobile couponing is a new opportunity for mobile operators, brands, and retailers. Yankee Group analyst Jon Paisner acknowledges that there are battles still looming over the mobile coupon ecosystem, but predicts that volume will emerge in the next few years. "There are some technology knots to untangle and some business and distribution issues to sort out. But mobile coupons provide brands and retailers the ability to create personalized campaigns to increase revenue while decreasing costs versus other marketing methods, even other mobile programs," explains Paisner.

―――

To make life simple for his tourist clientele, Habib takes mobile payments. Many tourists have not yet converted their cash by the time they stop by for their first street snack on the way to the Gateway of India, but it doesn't matter. All they have to do is swipe their mobile handset next to his payment receiver from mobile operator Bharti Airtel, and they're charged through their mobile bill. It also puts a stop to the petty thieves who used to hang around his father's cart, eyeing the cash and looking for the right chance to run off with it.

A perfect example of a future entrepreneur of the developing world, Habib plans to expand his business further. His long-term goal is to get some friends to operate carts of their own, with his twenty-first century samosa catering operation supporting them. He found a free download for his phone that will let him see his sales in real time, helping him spot trends and adjust his menu and pricing by location, time of day, or season of the year. He's also thinking that he'll put Wi-Fi Webcams on the carts—just to keep an eye on everything from wherever he is.

Connectivity, for Habib, is an opportunity to build a business and an exciting future of better prospects for him and his family.

Fatuma, the Kenyan Coffee Farmer

Growing coffee in Embu, at the foot of Mount Kenya, has never constituted an easy life for those who do it. For Fatuma and her husband

Juma, however, hard work is simply one of life's demands they meet as best they can.

Kenya is, in fact, one of the bright spots of sub-Saharan Africa, with both a relatively stable government and fairly robust economy. It continues to be the primary communication and financial hub of East Africa.

Life is not easy for Fatuma and Juma, but they are happy. Their first child is due in a few months.

But Fatuma worries about Juma's tuberculosis. It leaves him weak and sometimes unable to work, which creates strain for Fatuma. He needs regular treatment and medications. It's hard to get those kinds of things in a place so remote.

Tuberculosis is one of the largest health challenges in coastal Africa, second only to HIV/AIDS. Ensuring regular delivery of drugs can be challenging, meaning that clinics frequently run out of stock. Without basic information about patients such as their addresses and reliable ways to reach them, clinics can't get in touch with patients to save them a lengthy and fruitless trip, nor keep tabs on how consistently they are taking their medications. Connectivity via mobile phones and basic computers has begun to do a lot to improve this situation.

But when Juma traveled to Nairobi recently for treatment, the doctors there helped him work with Safaricom to get a cell phone. It's powered by solar cells; so as long as he doesn't forget to take it out of his pants pocket, it's always ready to use. It's all part of a program sponsored by the World Health Organization, helping Juma to coordinate with community health-care workers in Embu, who also have phones.

The health-care workers can call Juma to check up on him and make certain that he takes his medications. They can also order refills for his prescriptions. When he used to run out of ethambutol, it could take weeks to get a refill. Now with his phone, he can call the health-care workers in advance, and they can let him know when they have it in stock.

That's not all Fatuma and her husband can do with communications technology like their new mobile phone. Although neither Fatuma nor her husband can read well, they are still able to use a computer in the local market, hooked up to a car battery, to sell their entire coffee crop before it's harvested.

By estimating how much coffee they are likely to produce and stating when it will be ready, Juma is able to receive electronic bids from coffee buyers around the world. He and Fatuma receive some of the money in advance and use it to cover their immediate needs for medicine and preparing for the baby. Getting better prices for their most recent crop means they can buy a few supplies they didn't have for their first child. But it helps a lot that Juma's brother, working in Europe, is able to transfer some of his earnings directly to Juma for his nephew-to-be. Juma can get the cash from that transfer by going to the local storekeeper.

Connectivity even helps Juma and Fatuma keep an eye on their baby before it's born. Using a wireless sonogram wand, Fatuma can send a sonogram image to a health-care worker by phone. The health-care worker reviews the image, figures out the baby's due date, and scans for potential complications like a breech birth.

The ultrasound technology described in this scenario is being tested by the global nonprofit organization Partners in Health in conjunction with Massachusetts General Hospital. Today's units work with PCs, but simpler units are being tested for use with mobile phones.

Although Anywhere connectivity is coming to Africa more slowly than it is in the rest of the world, in a decade the change there has been phenomenal. The benefits of connectivity, even at the earliest stages of the revolution, are also incredible. The improvement to basic health care and the enhancement of people's ability to support themselves in some of the world's remotest regions make connectivity nothing short of life-saving for millions of people like Fatuma, her husband, and their baby.

Sheryl, the American Advertising Executive

Although it surprises her, Sheryl has found herself becoming one of those people who's happy if her kids are happy. She's worked hard to establish a lifestyle that lets her put her kids first, whether it's about going to their sports events or just sitting together at home to watch a movie. (Although sometimes the movie-watching gets interrupted a lot. At least she can pick up watching it from any of the screens in the house.)

Sheryl's father was a pharmaceutical sales manager who traveled all the time (it seemed like "all the time" to a 14-year-old kid with an attitude). Her mother balanced child care with part-time work at a local hospital. That took her away most evenings.

Sheryl now has three children of her own and wants the same things for them as her parents wanted for her. She works in advertising full time, which used to be an intensely demanding career, at least timewise. These days, though, she works from home three days a week or more.

At the back of her home, she has a pretty nice office setup. Walk through the door and you could almost feel as if you're stepping into her company's corporate offices downtown. On one wall, high-resolution screens give Sheryl a constant view of her calendar, agenda, notes, and to-do list, plus three clock faces showing the time in New York, Berlin, and Chennai. She also has a video link to her assistant and her creative team, whose work she can review and revise through a live collaboration application where they can all draw on the designer's proposals.

On the shelf behind her desk, she has just a few paper files; she got rid of her last filing cabinet last year when she realized that she didn't need to keep paper copies of anything any longer, now that the company's entire archive is online.

Even when she is working from home, Sheryl feels totally connected to her colleagues and her clients. It's all thanks to her superfast broadband connection. Her neighbors enjoy lower-cost broadband service, and one of them even gets it for free by accepting advertising with it, but Sheryl uses a premium service to her house that ensures she gets great network performance for her telepresence and collaboration sessions. It means that her network experience is great, even when the neighbors' kids are clogging up the network with their multiplayer battle games.

She can access all of her office resources online, and quite frankly has become addicted to telepresence. Working from home means that she gets to be fully connected to work but that she is still able to get dinner ready for the kids when they get in from soccer practice. It's just a 10-second commute down the hall.

The funny thing is that she can hardly tell the difference anymore between going into the office and working from home. It's true that at the office she no longer has her own space since it wasn't getting used much. Instead, she uses an online form to let the office manager know she's going to need a desk, and she gets an alert on her mobile as she drives in to tell her where she can set up on arrival. But once she's at the desk, all her files, calendars, and projects appear on the screens there just as if she were at her home desk. Even the photos of her family are there on the wall. Lots of employees have stopped going into the office with any regularity nowadays, but she still likes the in-person contact and ensures that the local members of her team meet up for lunch regularly.

When the members of her family are not together, technology helps them stay connected as well. Her cell phone lets her see where the kids are when she needs to (it's a lot less embarrassing than having a parent call to check on them every half-hour). But the center of action in the home is the same as it's always been—the hall by the back door where the kids dump their gear. She recently got a wireless screen mounted on the wall by the coat hooks that has a handy mirror in the middle, but its big value is that it pops the GPS-supplied locations of the family hand sets and cars onto a map that also can be overlaid with traffic and weather data. She can even locate their dog when he runs through the neighborhood. Her husband wants to add some sports scores to the display, and maybe he will; it's easy enough to change from any computer in the house. She likes the home energy consumption view, letting her spot at a glance whether they're keeping their electricity bill under control. There's not much to do. The machines mostly turn themselves off and on according to when the rates change, but it makes her feel good to know she's doing her part.

As for the kids, well, they love the bank's money transfer service with automatic alerts when their e-cash balances run low. There's not much cash in this household anymore, and it's proven to be a great way to teach money management. The kids not only have a budget, but they also know that they can get access to their parents' money in an emergency. It's a nice safety net.

The kids are addicted to video calling. Wherever they are, they almost always choose to video-call their parents from their cell phones. There's even a system to set up a quick virtual family meeting when needed. The video quality is so good that it's like everyone is in the same room together.

The sophistication of video calling is a major comfort. Sheryl's oldest daughter, Rachel, is leaving for college next year and plans to take a semester to study in Rome. Ten years ago, the thought of her daughter traveling so far away, by herself, would definitely have caused some sleepless nights, but Rachel is going to stay with her school friends, friends who just happen to live in Rome. She talks to them almost every day and is closer to them than most of the kids from the neighborhood. Sheryl has also been on a video tour of the university in Rome and sat down to a virtual dinner with the host family.

Webcams give her a peek at who's in the house when she's out; the other day she spotted Rachel having some older kids over that Sheryl didn't recognize. Of course Rachel had a fit about the invasion of her privacy when asked about it later; it reminded Sheryl of the battles she and her mother had over where she went and what she did. Since then, Sheryl has been asking other parents how they balance security and privacy concerns for their teenagers. Some things never change, even when the tools do.

For Sheryl, the biggest benefit of technology is that it allows her to have a fulfilling career without sacrificing family time. In fact, she can spend all the time she wants to with her family and still earn a living, and she gets to make the most of every minute. When the kids were little, it was so easy to create photos and videos that she could treasure every experience without having to think twice—no more fishing around in purses for cameras or videotapes.

Now that the kids are older, Sheryl can give them the space they want without missing out on the experiences they're prepared to share.

Whenever there's an important event happening for the kids, it's on the shared schedule in the network, and she and her husband get reminder alerts on whatever devices they're using.

With the kids growing up, though, Sheryl has more time to pursue some of her personal interests. Last year, she helped create a local community group that's taken over some of the roles previously assigned to local government. Members of this group stay closely connected through online facilities (Web site, blog, community networking site, mobile notifications), sharing common goals and working together to make the community a better place to live.

Community groups like Sheryl's do fund-raising and campaigning with great efficiency thanks to the network. They use the Web to debate, nominate, and vote on the projects that should be initiated to improve the area. They can also collect donations to make some of the necessary improvements by using secure money-transfer technology.

During one of the meetings last week, the elected official talked about how individuals will be able to vote remotely in the upcoming presidential elections. That will mean a lot less running around trying to transport people to the polling booth on Election Day.

For Sheryl, it's just nice to be able to take part in so many different activities without being overwhelmed by the demands on her time.

Arne, Retired Swedish Professor

In a cozy but elegant apartment in Stockholm, we find Arne, 67. For most of his adult life, Arne was a professor of music at the University of Stockholm. He retired two years ago and has since been enjoying the extra time to devote to his current hobbies: the arts and horticulture.

Swedish society is remarkable in a number of ways, but he is especially thankful for the robust information systems and governmental social programs that allow him to focus on the more enjoyable and productive parts of his life.

One of the least enjoyable aspects of growing older is the decline of health and energy. Arne receives a lot of attention from his physicians and their support staff to stay current with his healthy heart program and diabetes maintenance.

In the early days of his diabetes, he had to prick his fingertips to test his blood. Five years ago, the medical geniuses at the local hospital introduced a fabulous new product that was inserted under his left arm which regularly monitors his blood insulin levels as well as his cholesterol and triglyceride levels.

The tiny device does all the testing automatically and then transmits the results to Arne and his doctors over the Internet. The monitoring implant is wirelessly connected to an insulin pump, so that for the most part his diabetic condition is automatically maintained by this combination of devices. The regular stream of information coming from the device informs his physician about his nutrients levels and blood components, which are then analyzed on-site at the University Hospital. For Arne, it's no longer a question of remembering to check, because a special light in each room of his apartment glows red if his blood sugar gets too high. He can glance at it easily wherever he is, and he spends a lot less time worrying about it. Even with his failing eyesight, he always knows when his blood sugar is a problem.

Arne spends a lot less time at his doctor's office, mostly because he takes advantage of the home-based health maintenance programs. He makes great use of the personalized food service that is a feature of his health-care plan. The service responds to his specific nutritional needs on a daily and weekly basis. Based on the reports from his implant, a local food service delivers prepared food to his apartment twice a week. Arne can let the service know of any special preferences he has. It is able to quickly establish the inventory of food that he has at his home already through the **RFID (radio-frequency identification)** tags on the items in his kitchen. It can create daily menus and recipes for him, incorporating the foods he has available. The food service company can also initiate orders from local markets, keeping in mind what will benefit his diabetes and heart condition. Worries about food safety, so rampant years ago, have diminished for Arne, since most of what is delivered to his home has been tracked with wireless sensors from the point of manufacture all the way through delivery. The delivery service knows that everything is fresh and safe.

In combination with the health data from his implant, Arne also has a specialized watch to monitor his physical exertion levels and heart health. Every morning, he gets customized exercise routines sent to his mobile device to help set up regular exercise goals that are compared to his actual performance.

Getting out on the town is still very important to Arne, and a lot of the location information he gets from his mobile devices fits into the exercise program and incorporates all of his activity levels over the course of the day: whether it's walking to the store, working in his greenhouse, or getting out for a late afternoon stroll through town. The map feature on his phone gives him audible turn-by-turn directions wherever he is in the city. It's a great help given his less-than-perfect eyesight.

In 2008, Stockholm ran its first pilot test for a total digital inclusion project, designed to support people with visual disabilities. The project enables the visually impaired to walk around using a mobile phone that provides turn-by-turn voice navigation. "With every tree and every roadblock documented, you could let go of the guide dog," says Tomas Bennich, president of the Sweden Mobile Association. Using this navigation system, the visually impaired are able to walk around freely, saving time and money while enhancing their independence.

Connectivity is benefiting just about every part of Arne's life at this point. Not only are his health problems well-managed thanks to connected devices, but he can also manage most aspects of his life via his mobile phone. He can project the display from his phone onto any surface he wants, meaning he doesn't need to squint to read it. He can plug the device in at home and watch TV or browse the Web from his phone. It's also a backup safety device, since it picks up the signals from his implant and sets off an alarm if his blood sugar level rises or if there are any other medical complications. Of course, the alarm is just to alert anyone who might be nearby and able to help; the device also sends an alert, complete with GPS coordinates and Arne's vital signs, that will bring the nearest emergency service personnel to the scene.

When he uses public transportation, he doesn't need to worry about trying to figure out his fare. As long as he has his mobile phone, the rail system understands his location and automatically processes whatever he owes, applying his senior discount before deducting the total amount of his ride from his bank account.

Arne can also stay active in the music community using Anywhere technology. Via the fiber-optic Internet service in his home, he can work in real time with a lot of his musical friends and influences, not to mention rehearse with his baroque chamber music quartet, which holds live rehearsals from multiple locations and records the session for future mixing. The group is able to book new performances and accept performance requests on its dynamic Web site, not to mention watch many of the outstanding baroque groups from around the world through high-definition concert footage available on demand. With 3-D projection and surround sound, it can often feel as if you're there in person.

Seeing the Anywhere Consumer Opportunity

Despite the obvious differences between an Indian street vendor, a couple of Kenyan coffee farmers, a white-collar working mother of three, and a retired Swedish professor, these people do have a common denominator: lives transformed by embracing the connected experience. In a relatively short time, we see increased economic vitality, better health care, greater life expectancy, and better education—all thanks to connectivity.

For business leaders, there are also some revolutionary opportunities for commercial benefit embedded in these profiles. A few stand out:

First, these consumers have many *more things* in their lives. You probably spotted some of the new electronics that our more well-off consumers talked about: entertainment devices, kitchen appliances, personal care tools, and more. Some of these may be new product opportunities for you, but both our teenaged street merchant and the coffee farmers will be greater consumers of *all* the world's goods than their compatriots in emerging markets today. In our scenarios, they have mobile hand sets, payment technology, and other connected tools—all of these items represent increased profit opportunity for businesses.

Multiple studies exploring the world's emerging markets have linked measurable growth in GDP to the rise of wireless networks in the regions. Rajeev Suri, CEO of Nokia Siemens Networks, a leading global enabler of telecommunications services, talks about the impact he sees in markets like India. "As the saying goes, 'it's expensive to be poor,'" he says. "Poor people often pay more for the same product than rich people do; lack of choice, lack of competition, and physical distance all contribute to the problem. Being connected helps to break those inefficiencies. When you can spend less on the things you need—water, food—you can spend more on other things." Can your company expand its product offerings and global reach to be in place when consumers in emerging markets can begin to afford a new class of goods?

Second, *commercial payment methods* will look very different across the board and will create market-entry opportunities for new companies. Where consumers don't have access to conventional banks and credit, they may never develop the habit of using them if the first noncash solutions they're offered come from other sources. Mass-market brands that these people already trust—network providers like India's Bharti Airtel, retailers like Subhishka, and other firms connected in the consumer's mind with money or transactions—may be able to develop that transactional relationship with the consumer instead.

Western Union, the global funds transfer company, has moved swiftly to capitalize on its brand and existing infrastructure in expanding financial services in emerging markets with high mobile penetration. Matt Dill, who leads the firm's digital ventures, says, "Western Union serves a group of people who in the past have been disenfranchised. We've created a central brand and a trust network of locations where we have a contract with a local agent to fund a cash payout from a transfer done by mobile phone. We're able to offer a more efficient product to the underbanked, or unbanked, than you or I have when doing funds transfer today. It's the transactional channel of the future." (See Figure 5.2.)

A third commercial theme ahead is the *expansion of services* to consumers. It's hard to provide services to people who have to walk

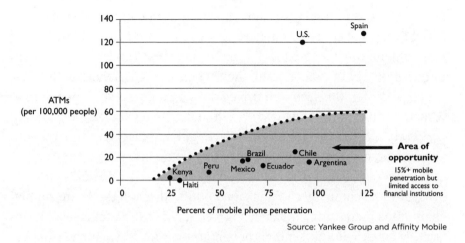

Source: Yankee Group and Affinity Mobile

Figure 5.2 *Mobile phones open up new financial opportunities.*

miles to get to you or who are too infirm to leave their homes to visit your clinic or take in a concert. But when the network has the capacity to deliver complex, multimedia experiences to more devices than just a PC in places we spend time whether that's home or elsewhere, businesses can deliver more services. What products does your firm offer to the consumer that could be linked to you to provide an ongoing service?

Fourth, a more connected future is one with *new distribution solutions*—getting things to buyers in new ways. When retailers can place orders on the spot, getting product to them can be less expensive, since less of it needs to wait in a warehouse or rot on a dock lacking refrigeration.

Nortura, a Norwegian food manufacturer, is already pioneering the use of inexpensive RFID sensors that are applied to food product packaging so that the location of the food can be constantly tracked. Are Bergquist, CEO of Matiq, the division the company launched to pioneer the technology in its food processing centers, explains the benefit: "Nortura needs to unload trucks arriving from other centers and reload them to go out to our customers, the grocery stores. Checking in stock in that process, to make sure the right things get on the truck, has been done manually and it takes hours. While that's being done, the products aren't available to be ordered yet by a customer because they're technically still in transit and not entered into the distribution center's system yet. That could mean a lost sale even though we have the product right there. But if we affix each pallet with an RFID tag with information about its contents, then when a truck arrives at the distribution center, the second it drives through a portal equipped with a reader, its precise contents will be automatically registered. We can sell the product instantly, and we have accurate management reports and virtually no errors in packing trucks for customers."

Using a global standard for encoding product information electronically, the company can track its products through the entire supply chain, thus increasing efficiency and reducing cost. Retailers can keep track of stock without sending people to look for it, saving expensive labor in a traditionally low-margin business.

Media distribution shifts even more rapidly in an Anywhere world, since the product travels on the network itself, allowing new firms to enter the market with a unique perspective on how to reaggregate content for the fast-changing appetites of small groups of users who were previously not viable to serve for economic reasons.

Finally, if you had your marketing hat on, I hope you spotted the wealth of potential opportunities to create *more awareness and consideration* of your product or service in the Anywhere futures we described. Whether it's coupons, or advertisements on the mobile, the connected device that's always in a pocket or purse is a tremendously useful new screen for marketers.

SampleSaint is an innovative marketing firm launched in the United States in 2009 around mobile couponing. Its technology provides images of on-sale items that consumers can download into their phones either from a Web site or in a store, to be scanned at a grocery checkout stand for a one-time discount. Global consumer goods giant Unilever has been experimenting with the technology in partnership with grocery store chains. Lawrence Griffith, CEO of Samplesaint, points out the commercial value that mobile couponing offers in several areas. "Mobile couponing is cheaper; you don't have the cost of the media buy, and you save the printing expense, the retailer costs, and the clearinghouse charges. This is just a direct savings at the cash register. And our early experiences are that redemption—how often the coupon is actually used to purchase a product—is over 25 times higher. So executives are seeing that it drives purchase." But the biggest profit payoff may come from eliminating coupon fraud. Just one recent consumer coupon fraud case in Milwaukee, for example, alleged over $250 million in fraudulent savings.

Young & Rubicam North America leader Tom Sebok notes that couponing was originally designed by brands primarily to attract new customers to try products for the first time and to reactivate lapsed or infrequent users. "But given the intensely personal relationship people have with their mobile devices, mobile coupons can be expected to work differently. For example, they might not attract a lot of new customers. Nonetheless I think it's an area that's bound to grow because it will be especially useful for retailers. Their ability to influence a current customer's frequency of visits as well as the contents of his or her shopping cart is very high. Say you shop at a do-it-yourself retailer like Lowe's. Because you're a regular customer, you probably won't mind getting a mobile message when there's a sale on fertilizer. It'll be a welcome enhancement to your relationship with Lowe's, and they can manage it profitably because they'll be able to increase your frequency and your spend with relevant offers tailored to you," he predicts. (See Figure 5.3.)

Connected devices could become so inexpensive, and at the same time so valuable a vehicle for a consumer dialogue, that mass-market consumer brands—retailers, consumer goods manufacturers, media

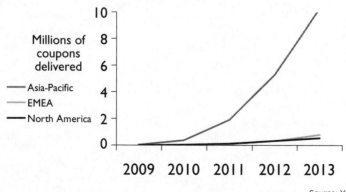

Source: Yankee Group

Figure 5.3 *Forecast for mobile coupons delivered annually.*

outlets—will eventually distribute the devices themselves to gain the opportunity. It's not such a stretch; in multiple markets around the world mobile operators already do this. That is, they subsidize the retail price of a mobile phone for consumers to ensure that they capture the longer-term subscription relationship with them, which is more valuable than the one-time profit to be had from selling the phone.

Consumer device entrepreneur David Rose of Vitality says, "There will be many more 'screens' in the home—not just conventional screens, but more surfaces that could be used to communicate with us." Chumby Industries' Steve Tomlin agrees. "The next compelling venues in the home for devices are beyond the living room. The kitchen is an important one, but there are opportunities in every room in the house." How will your company use the explosion of new pathways to the consumer to create awareness and consideration of your offerings?

Ten years from now, ubiquitous connectivity will have brought substantial change to consumers all over the world. But we're already beginning to scratch the Anywhere itch—expecting to take our experiences with us anywhere we go.

In the next chapter, we explore our current behavior as consumers and see how some of us are starting to change into Anywhere Consumers. Early adopters of Anywhere Experiences are leading the way, helping to change the rest of us in the bargain. While you prepare for the consumer of 2020, you need to respond to the Anywhere appetites of today.

Chapter 6

Our ANYWHERE Appetites

"We want what we want,
when we want it.
We want to be
in control."

—Larry Weber, chairman and CEO, W2 Group

In the first two chapters of this part about the Anywhere Revolution and its impact on consumers, we examine some of the driving forces that will spawn more connected devices, and then we make a visit to the future to see how those forces and our varying situations will create new experiences for us all. But admittedly, the world doesn't have fully ubiquitous connectivity yet, and the majority of us have not yet become true Anywhere Consumers. As businesspeople concerned for the most part with making money next quarter and the one after that, do we need to care yet?

The answer is yes. Not just because we should be looking forward to and planning for the future, but because the revolution creating the Anywhere Consumer is measurable. Just as we use the Anywhere Index to compute the connectedness of countries, we can similarly gauge the connectivity interest of consumers. The tribes of consumers formed by these metrics will in turn yield some useful insights into how businesses can and should reach them.

In this chapter, I show you that some consumers are beginning to exhibit Anywhere attitudes and behaviors now, meaning that you should be adapting your approach today. I'll do this by answering three questions:

1. Who is Anywhere today?

2. Which consumers are the most influential in the Anywhere Revolution?

3. How can you target and profit from Anywhere Consumers?

We start that process with a quick review of concepts from Marketing 101.

Who Is Anywhere Today?

Segmenting consumers has been a part of consumer marketing strategies since the 1930s, when Alfred P. Sloan used traditional demographics such as age, gender, and income to target different brands of General Motors automobiles. Using its motto of "A car to every purse and purpose," GM sold Chevrolets to middle-class families, Pontiacs and Buicks to upwardly mobile men, and top-of-the-line Cadillacs to executives wanting to show they had "made it." Consumer segments allowed GM marketers to sell products to entire classes of consumers efficiently, while excluding other groups for whom those products were unaffordable or otherwise unsuitable.

With the evolution of the PC in the 1980s and 1990s, technology companies like Lotus and Microsoft redefined these segments to target technology buyers effectively. Now the willingness to buy at high prices and keep up with a constant stream of technology updates defined the new top consumer segment. While these new alpha adopters were college-educated white males with high disposable incomes, those traditional metrics weren't enough to identify early PC adopters. Marketers had to add another factor—technology interest—to their segmentation metrics to clearly target the most influential buyers of PCs.

The Anywhere Revolution is changing segmentation criteria again. When we at Yankee Group analyzed data we collected in our surveys of North American and European consumers in 2008, we found that the historically accepted early adopter—a young white male with high disposable income—is *not* the face of the most influential group of Anywhere Consumers. Instead, we found two behavioral attributes that were more predictive of consumers being influential adopters of connectivity in North America and Western Europe. Those two attributes were:

1. **Digital media interest:** How much we enjoy digital content such as music, video, e-mail, and TV.

2. **Connected device ownership:** How many network-connected devices we buy and own to participate in the Anywhere Network, be they computers, laptops, cell phones, networks, or iPods.

These two factors are sufficient to divide consumers into five segments that exhibit markedly different behaviors and attitudes about Anywhere, as you'll see in Figure 6.1. They allow us to measure how many consumers fall into each category.

From where I sit, this connectivity-driven analysis provides a pretty important breakthrough. It helps businesses focused on Anywhere products and services for consumers—from the most to the least con-

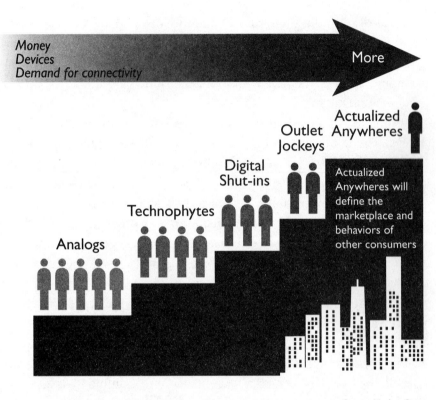

Source: Yankee Group

Figure 6.1 *The five segments of Anywhere Consumers.*

nected among us. Even among consumers who buy and use the same connectivity product, distinct patterns emerge about why we decided to buy it, or what we actually do with it. This means that marketers can offer the same product for sale to multiple Anywhere groups, but they should be adapting their messaging to respond to different needs.

To help clarify each of these distinct consumer segments, we've named the five groups to create intuitive impressions of each segment and how to reach it. Those names are the Analogs, the Technophytes, the Digital Shut-ins, the Outlet Jockeys, and the Actualized Anywheres. First I introduce you to each group in detail, exploring its most prominent characteristics. Then we tackle how to adapt Anywhere product and service messages to each group to maximize the profitability of your efforts. (See Figure 6.2.)

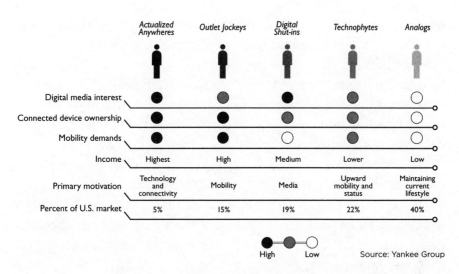

Figure 6.2 *Characteristics of Anywhere Consumer segments.*

1. Analogs

In a broad population, inevitably some people just aren't interested in new technology advancements. In previous technology revolutions, these were the people who thought electricity was a frill or that cars would never replace the horse and buggy. In this revolution, we've named these people Analogs, and the best way to describe them is to say that they are uninterested in digital technology. These are the people with VCRs still flashing "12:00" all day long. Analogs cannot even be called slow followers of our Anywhere trend, since they make the choice to avoid advanced technology altogether.

Much of their reluctance may be the result of age; the Analogs are, in terms of average age, the oldest segment. Because of the Baby Boom demographics in North America, the Analogs are also our largest segment, comprising 40 percent of the population. In Western Europe, the size of this group is even higher, at 54 percent of the total population. As today's new connectivity becomes commonplace for a younger generation, some of them will eventually "age into" this group, having adopted some basic technologies like digital music and online newspapers, but perhaps resisting later advancements such as Twitter. So while the group might diminish in size over time, we can expect that a core of connectivity-averse consumers will remain nearly indefinitely. It's what they will be averse to that will change.

Members of our Analogs segment tend to have the lowest household income and therefore the least disposable income. They are very price-sensitive and unlikely to opt for expensive advanced services such as high-definition digital video recorders (HD-DVRs) or mobile data services for their cellphones.

Although age plays a major role in determining the attitudes of members of this segment, their basic attitude toward technology is key. They are simply not interested in advanced technology and don't buy technology for technology's sake. Slowly technology will be adopted by this segment, but only when it makes existing activities easier. For example, Analogs may be quite pleased to use a digital photo frame that automatically updates with new images of grandchildren. They accept the new device because it fills a need for them and eliminates the need for another piece of technology, the PC.

But these instances will be the exception to the rule. Despite the Anywhere Revolution's reaching more than 5 billion people by 2015, this group will be the least touched by that revolution. Other segments seem as if they are born to a world of ubiquitous connectivity; for the Analogs, connectivity will instead be thrust upon them or ignored.

2. Technophytes

As the name suggests, Technophytes are a group of consumers who aspire to be considered cutting edge when it comes to technology. In reality, though, they are late followers of digital media and connected devices. Why? Because while they are interested in these technologies, they don't have the disposable income available to buy them until they are mature and more affordable. Technophytes want their Anywhere connections; they just can't afford them until they are mainstream.

Given their income restrictions, why are Technophytes so interested in Anywhere devices and media? It's simple: they see Anywhere technology as conveying status and upward mobility. When they can afford them, these consumers proudly show off advanced smart phones such as BlackBerrys and iPhones as badges of accomplishment. They view these devices as aspirational but often unaffordable paths to a more prosperous future. Technophytes' interest in technology assures manufacturers and service providers that consumer demand for advanced devices can extend beyond the early adopters. As product prices decline, Technophytes will provide a source of volume in sales. Therefore, marketers should look to this group to offset plateauing sales of relatively mature Anywhere products. That's because Technophytes are heavily influenced by Actualized Anywheres, Outlet Jockeys, and Digital Shut-ins, three groups we discuss shortly. Pleasing these other segments and gaining their adoption of connectivity solutions will win Technophytes in follow-on marketing.

3. Digital Shut-ins

Digital Shut-ins is the name Yankee Group has given to a group of consumers who have a great interest in digital media in their own homes, but less so when they are away from home. While these consumers enthusiastically embrace digital media, they aren't as aggressive in adopting connected devices. Educating this segment remains important because Shut-ins often fail to equip their devices, such as HDTVs, with all the services they support, like HD channel programming from the providers.

For these consumers, Anywhere means "any room in my house" rather than the mobile experiences many of us might assume. These consumers are rabid fiber-optic and cable TV users, who view watching a football game on an iPhone as weak tea compared to the big-screen experience of their flat-panel LCDs. Triple-play services that bundle broadband Internet and phone lines to their monthly pay TV bills are the types of product that appeal to Digital Shut-ins.

Thus Shut-ins represent an important segment for in-home Anywhere products and services, some of which may be untapped by current markets because marketers don't see the full potential available in the attitudes of these consumers. The portable lifestyle beyond the home is less important to them (they are less likely to own a digital camera or digital camcorder, and are also uninterested in buying portable devices in the near future). But they clearly care about their at-home experience. In 2009 North American Shut-ins spent an average of $84 on their monthly television service to make it pleasurable.

This segment is probably the single most lucrative target for cable and fiber optics companies, simply because Digital Shut-ins want the premium wired experiences those providers offer and don't care about the mobile offerings of their competitors.

4. Outlet Jockeys

Mobility is the watchword of Outlet Jockeys. If they had a motto, it would be "Why go home?" Occupying a position diametrically opposite that of Digital Shut-ins, Outlet Jockeys cherish advanced, out-of-home devices and mobile services more than at-home wired experiences. You can easily identify Outlet Jockeys in public; because of their battery-powered gadgets, these people are usually hunched over the last available power outlet in the airport or hotel lobby.

Along with consumers in our final segment, Outlet Jockeys own the latest and greatest portable devices and thirst for more of them. They want satellite radio, digital camcorders, and mobile Internet devices too, and they are great targets for new technology such as geo-tagging, Wi-Fi memory cards, and connected digital audio players (DAPs). These mobile consumers enjoy cutting-edge portable applications as well as using their hand sets to e-mail, instant message, and engage in social networking.

Despite owning do-it-all **smartphones** such as the iPhone or the Palm Pre, this group brings laptops, digital cameras, and MP3 players on the road with them. Their behaviors indicate that one Swiss army knife device will not replace multiple individual best-of-breed products. Businesses will have to consider this when they design Anywhere products.

Perhaps the most interesting fact about this group is that being an Outlet Jockey is simply a way station to other things. When their need for travel and mobility wanes, Outlet Jockeys bring many of their Anywhere behaviors home with them. As a result of this behavioral transference, Outlet Jockeys are most likely to evolve into our final and most connected consumer segment, the Actualized Anywheres.

5. Actualized Anywheres

Actualized Anywheres may be the smallest segment, comprising only 5 percent of the population in North America and Western Europe, but they are the most important. They enthusiastically buy Anywhere products and services for their mobile lifestyles, and, unlike Outlet Jockeys, they apply those devices and services at home. Actualized Anywheres bring the concept of a ubiquitously connected consumer to life.

This group's consumers punch above their weight in market impact, in part because they spend the most on their connected devices and services, buying before economies of scale help to lower prices. This group also is the trailblazer for every other segment, wielding huge influence on the general population. In a world in which an increasing number of consumers are relying on the feedback and advice of their peers, consumers who adopt first dramatically affect success in the broader market. Companies must make seamless connectivity core to the experience they sell to please this network-conscious segment.

This group undermines the stereotypes many in the technology industry may have about the typical first users of technologies. Rather than being composed predominantly of white, male, mid-level 20-somethings, this group is actually racially diverse, has people with the highest level of household income, and encompasses a wide age range, between 18 and 44 years old.

Actualized Anywheres pay the most for the services they subscribe to, but they are loyal to no one brand or company. They buy from whatever provider meets their Anywhere needs most effectively, without the slightest remorse at abandoning old brand relationships. They also buy through various channels and from a multitude of man-ufacturers. Remember our Zagat versus Yelp story from Chapter 4? Actualized Anywheres help crown these new Anywhere leaders as they seek the best connected experiences. Marketers should not assume that Anywhere Consumers will stick with their brand.

Which Consumers Are the Most Influential?

We have now looked at the five different types of Anywhere Consumers in our world. On a walk through any typical downtown or city center, you will most likely spot an Analog, a Technophyte, or an Outlet Jockey among the people you meet. Based on the characteristics we mentioned, you should also be able to pinpoint your own Anywhere style. Are you a Technophyte? Perhaps you're an Actualized Anywhere?

Anywhere Banking: How Bank of America Moved into Mobile

As we begin to evolve into Anywhere Consumers, wanting to take our experiences and digital activities with us wherever we go, brands that want to stay connected with customers must keep up. Doug Brown, senior vice president for mobile product development at U.S. retail financial services leader Bank of America, talks about customer appetites for Anywhere banking, what the firm has implemented so far, and what may lie ahead.

Q. *What made you decide to act?*

A. We noticed in 2006 the uptick in behavior of our customers with mobile. It looked like it could be more powerful than the online channel.

We began to assess the demand for it from customers, and all the primary and secondary research confirmed that. We also kept our eyes on technology enablers making this more prominent, more likely to happen. We decided the time was right in early 2007 to launch a mobile banking platform.

Q. *What did you do?*

A. We didn't want to be limited to mobile phones only. We were very aware of other handheld connected devices, like the Sony Mylo, the Apple iPod Touch, and a whole new set of devices that will follow them . . . so clearly it's not just a phone game, it's a mobile lifestyle. We went with a mobile Web implementation, so that any connected device with a browser could use the mobile experience at Bank of America.

On our platform, mobile customers can perform a variety of banking functions, including checking account balances, reviewing transaction details, verifying payroll deposits into accounts, paying bills, and transferring funds to other Bank of America customers.

When we launched the program, it wasn't just a product launch. We had to teach, educate, and bring our customers along with us.

Q. *What kind of customer reaction did you get?*

A. We expected that the student population would be the early adopters and that they'd refer us to other students. That did start to play out. But what was amazing is how they brought their parents and grandparents into the service. They said things like, "Hey Mom, I'm in the bookstore and I need that money now, so just open the browser on your phone to transfer it." The network effect, each person's microeconomy, helped it take off faster.

As more people began to use it, they began to find use cases we hadn't thought of. As we expected, they would do payroll deposits, check to see if a deposit cleared, and pay a few bills from the mobile channel.

But at the retail point of sale, they also check their balance right as they step up to the register. We have learned that it influences their spending when they have that immediate access.

But it's not just transactions that customers are interested in. It's counsel, too. Basically their message to us is to make them more constantly aware. They say things to us like, "Please coach me, please help me stay on my goals." "Track my budget." "Tell me whether I'm getting a good deal on this purchase." And with the advancements in mobile, we can be with them all the time to do that.

Q. *What kind of impact did it have for Bank of America?*

A. As of August 2009, we had over 3 million active users, which was about more than 10 percent of our online banking base at the time. Mobile has directionally supported where we want to be, because it's what our customers want. We continue to expand and provide more capabilities.

Q. *What lessons have you learned so far?*

A. Consumers want entire capabilities in their mobile devices. There is an insatiable demand; they just expect more. I think they will expect their mobile devices to access their accounts, pay bills, and do transaction lookups. But they'll also want it to be their point-of-sale instrument, replacing all their plastic. To me, that means that the mobile device becomes the payment instrument itself, part of a totally connected commercial experience.

Q. *What advice do you have for others who want to leverage similar technology for their business?*

A. There are incredible things yet to be exploited from a range of innovations in mobile technology. Biometric authentication can be layered into the experience, for example; but you have to balance those developments with the customer's readiness. There are big social challenges for them to adapt to.

The connected world is a very fast, rapidly emerging ecosystem. It's emerging; you have to go with the flow and take chances. You have to have a general direction to head toward and then try multiple things to address some possibilities of how it will evolve. You don't get it right by picking one approach and expecting 100 percent success. You need to plan to do a couple of things, where some things just won't be as successful as others, and then refine what works.

My big learning was: Do it faster. You can never do it fast enough for the mobile customer's expectations. Once you let your connected experience out of the box, the incredible insight and intuition of customers will take you on this journey from there.

Wherever you fit among these five groups, you must recognize that everyone has the potential and capacity to shift between groups, although not everyone will. Each group also has an impact on the others. This is particularly true of the Actualized Anywheres, the most advanced and most influential consumer segment of the five. So let's take a closer look at this group.

When Yankee Group first introduced the concept of Anywhere Consumers, we set out a few basic rules to identify them. First, we noted that Anywhere Consumers carry several intelligent connected devices, such as BlackBerrys, iPhones, Kindle and Reader eBooks, iPods, and digital cameras with wireless connections. Secondly, they have the most advanced technology in their homes. They have HD-TVs, DVRs, high-speed broadband connections with wireless capacity, and sophisticated and entertainment devices, many of which also have connectivity.

Actualized Anywheres use technology to get their entertainment wherever they are, whether it's side-loading podcasts from iTunes before they leave the house, using place-shifting technology like Sling-box to access their home TV content, or streaming music from network-based services like Pandora. Connectivity is at the core of their every activity. The greater their consumption of multimedia content and services, the more important the quality of service, speed, and network reliability become.

Actualized Anywheres are the pioneers of connected devices, but they also are the guinea pigs for those devices when they are early in the market and unrefined. Unfortunately for some device makers like Sony, bad early experiences can have devastating results on broad market adoption.

Sony was late to enter the digital audio player market in the United States because it continued promoting its CD players such as the Disc-man while the market changed around it. Sony believed it could extend its MiniDisc player success in Japan and Europe to North America, and as a result, it was slow to develop a hard-drive-based player that could compete with Apple's new iPod. Actualized Anywheres were the first to recognize Sony's shortcomings and adopt iPods from Apple instead. Sony's delays and these Actualized Anywhere defections proved disastrous for Sony; its operating profit for its 2005 fiscal year fell from 160 billion yen to 110 billion yen. Meanwhile, Apple sold more than 30 million iPods in that same year.

In 2004, *New York Times* reporter Ken Belson pointed out Sony's failure: "How Sony got outflanked is as much about Sony's inflexibility as Apple's initiative. With its ownership of premier music labels and its foundation in electronics, Sony had all the tools to create its own version of iPod long before Apple's product came to market in 2001. But Sony has long wrestled with how to build devices that let consumers download and copy music without undermining sales to the music labels or agreements with its artists."

Sony put its content relationships ahead of consumer interests in connected media. The Actualized Anywheres, who are the least loyal of our consumer segments, weren't about to wait for Sony to figure it out; they fled to a competitor who was willing to satisfy their need for instantly available digital downloads, and they took the rest of the market with them. Sony has yet to regain its success in the music market, all because it ignored the needs of this small but influential segment.

How Can You Target and Profit from Anywhere Consumers?

Would an Analog stand in line overnight for the first generation iPhone? Of course not, but Actualized Anywheres and Outlet Jockeys do. In doing so, they create a buzz that reverberates throughout the mass market.

Apple captured the interest of the two most advanced market segments because it created a product that afforded users freedom and usability unlike anything available to them before. Others can capture this buzz too, but they must find the right message for each segment. While Actualized Anywheres lead the Anywhere Consumer movement, 95 percent of the population still clamors for last year's hot connected device.

Companies such as INQ have followed Apple's lead in creating connected consumer products. Targeting a younger demographic than that served by the iPod, INQ released a new Facebook phone named the INQ1 in 2008. It was designed to be the ultimate social media tool with

Facebook, Skype, and Windows Live Messenger incorporated into the phone along with a wide variety of widgets. Targeting a younger audience than most smart phones, the INQ1 integrates the best Facebook features (such as messages, pokes, and friend requests) for maximum connectivity, all in a low-cost device. It has built on its success with follow-on hand sets, including a Twitter phone. INQ has shown that services that Anywhere Consumers love plus devices that provide Anywhere access to those services are a winning combination.

The same formula—combining consumer electronics devices with network services—can be applied across a spectrum of offerings. Pay TV operators can allow remote DVR access over the Internet. Content owners can deliver their digital media in formats like MP3 that can be played by as many devices as possible. Consumer electronics manufacturers can build connectivity into their devices. However, before these offerings can succeed, manufacturers must target the right message to the right consumer segment.

An example might help illustrate this process. Imagine that you are the manufacturer of a connected digital music player that takes voice commands and gives voice navigational directions to people on foot. The device is about the size of an iPod shuffle, it costs $79, and you've named it the vGuide. You'd like to market this device to Anywhere Consumers. Where should you start?

The first step in creating your marketing campaign is to identify which Anywhere segments are most appropriate for your product. Figure 6.3 uses answers to two behavioral questions to differentiate the five segments; we simply choose the segment whose answers are most representative of the people who would use our device. Our vGuide device would be a pretty good answer to the first question, "How would you find a new restaurant to try?", and it's a good match for the Actualized Anywheres and Outlet Jockeys who use their mobile phones for similar functions. So those two segments, representing 19 percent of the population, are our target market.

But we're not done yet. The themes associated with our target segments in this figure can help us craft the right message to each of these groups. When we appeal to Actualized Anywheres, we should highlight how the vGuide will increase their freedom while allowing them to stay

		Question 1: How would you find a new restaurant to try? **Question 2: How would you catch up on a missed TV show?**
Category	**Theme**	**Answers**
Actualized Anywhere	Always connected life	Q1. Use Yelp to get restaurant reviews Q2. Pick it up on Hulu.com or stream from friend's Slingbox
Outlet Jockeys	Bridging gap between home and mobile	Q1. Use Google Maps on my mobile phone Q2. Download and watch it on my iPod
Digital Shut-ins	Creating a better home media experience	Q1. Call a friend near a computer Q2. Watch it on TiVo if I didn't catch it live
Technophytes	Cutting-edge affordable technology	Q1. Text my friends for recommendations Q2. Put in Netflix queue or download using BitTorrent
Analogs	Make existing activities easier	Q1. Use guidebook or paper map Q2. Check *TV Guide* for next showing

Source: Yankee Group

Figure 6.3 *Identifying, understanding, and selling Anywhere.*

connected. When targeting the larger population of Outlet Jockeys, though, we'll want to emphasize how the vGuide bridges the gap between home and mobile environments. So we'll need two different sets of messages to address the two segments we want to use our product.

This tailoring of marketing messages to our targeted Anywhere Consumer segments isn't unique to our vGuide. In fact, Figure 6.4 shows how we might craft value propositions for five different consumer electronics products to match the underlying interests of the Anywhere Consumer segments. Why address all five segments when the target market may be just the top one or two? It's simple: For every Actualized Anywhere in the population, there are another 19 people who are in other segments. While the top segments are influential, any mass-market product will have to sell to all the other segments as well. And because the Anywhere interests of these segments are all different, the benefits of Anywhere devices that will compel them to buy will be different as well.

So we can now market a single product to multiple Anywhere Consumer segments. But how can we profit from our knowledge of those segments? Again, the answer is simple: We build new products specif-

	Connected eBook Reader	Mobile Internet device (MID)	Wireless photo frame
Actualized Anywheres	You can't buy a paperback when you're stuck on a bus	No need; your smartphone will do just fine	Marrying the on-the-go and at-home experiences
Outlet Jockeys	Fit hundreds of paperbacks in your pocket	Another toy— with Skype!	A device that can display the photos you automatically load to the Web
Digital Shut-ins	Bring your entire library anywhere	Check recipes online while in the kitchen	A new screen to display content on
Technophytes	The newest way to read books and newspapers	If a smartphone is too complex for you, a MID is more familiar	A nifty device to show off to your friends
Analogs	No need to clutter up your house with novels	Easier to move around than a laptop	See new pictures of your family without a wait

Source: Yankee Group

Figure 6.4 *Segments respond to different messages.*

ically to meet their needs and interests. What should those products be? Here's a high-level list to get you thinking:

1. **Actualized Anywheres need connectivity embedded in more devices.** The next leap for Actualized Anywheres is having more connected consumer electronics beyond today's PCs and mobile phones. Amazon's Kindle eBook Reader whetted this segment's appetite for connectivity by using book purchases to subsidize its

3G wireless broadband connection. Actualized Anywheres will now be looking to upgrade nearly every device they own to connected versions, from their air-conditioning to their clock radios. Manufacturers who can capitalize on these avant-garde connectivity tastes can charge high prices and reap higher profits.

2. **Outlet Jockeys need better bundles.** These mobility-conscious consumers want to bring Anywhere home with them; they just haven't had time. With nearly 20 percent of this segment not subscribing to bundled home services, there's a significant opportunity to up-sell these consumers with integrated advanced **triple-play** and **quad-play** services. And since consumers rarely cancel these "sticky" multiplay services, selling to Outlet Jockeys now can lock in revenue streams for years to come.

3. **Digital Shut-ins require cheaper smartphones.** Digital Shut-ins are willing to spend money on some services, but they need more advanced media-capable mobile phones to take their home environment Anywhere. Subsidized smartphone prices and family plans for data will go a long way toward moving this segment off the couch and out into the world—without abandoning the home media they enjoy.

4. **Technophytes need lower prices overall.** Technophytes want to be cutting-edge connectivity adopters more than any other segment, but they just don't have the money to buy the latest and greatest gear. Hardware subsidized by service revenue, rent-to-own, and **prepaid** service plans all are ways to appeal to the Anywhere desires of this segment without exceeding their ability to pay.

5. **Low- (or no-) cost hardware will appeal to Analogs.** The members of this segment have lived most of their lives without technology, so simple functionality at low or no cost is the best way to get this group's attention. However, Analogs will likely find themselves adopting connected technology through health-care initiatives because they are the oldest of our five segments. Products such as Vitality's connected pill bottle that tracks medication adherence have no buttons and yet deliver significant health benefits to both

the doctors and the patients. Best of all, such products are easy for health-care providers to subsidize because they have a definitive and measurable return on investment.

Can you get this right? No time like the present to get started. Don Tapscott, noted researcher on the relationships between the Internet and young people and author of two seminal books on the topic, *Growing Up Digital* and *Grown Up Digital*, calls the people born between 1977 and 1997 not the Millennials as other social scientists do, but the "Net Gen." "The year 2000 didn't really alter the experience of young people of that time," he says. More importantly, he points out, they are the first generation that has grown up surrounded by technology, "bathed in bits." These consumers, Tapscott emphasizes, have different expectations than their parents about technology-based products and activities: they want to have lots of choice, to be able to customize their experiences, to collaborate with the companies they buy from, to have fun with whatever they do, and to experience near-instant service. These appetites will only accelerate the development of Anywhere Consumers."

And the lessons that identify the appetites of the Anywhere Consumer ̖ ̖ ̖ ̖limited to connectivity services and mass-market consumer ̖ ̖ ̖ ̖ ̖ics products. Bank of America began pioneering its first Anywhere retail banking offerings in 2008. See how they approached the problem and what they learned in this chapter's sidebar.

This part of the book on the impact of the Anywhere Revolution on consumers has looked at how our devices will create more connected experiences for us, the many opportunities for profit that will accompany them, and how our lives will change as a result, meaning businesses who target us with connectivity services must adapt their offers to our varying desires.

But I promise that the advent of expanding connectivity in our world will have just as much, if not more, impact on the commercial world. It will change what we do in business and how we do it. The next part of this book looks at how that's going to happen. As this revolutionizes how we do business, you'll see a third major area of profit opportunity for businesses that get involved before we move on to mapping out your own course of action.

PART III

The ANYWHERE Enterprise

Chapter 7

The ANYWHERE Consumer Goes to Work

"Work is the new 'killer app' for computing."

—Maynard Webb, chairman and CEO, LiveOps

As connectivity moves further into our lives, the divide between work and home becomes more difficult to distinguish. In the past, our professional and personal lives were separate and very obviously so. Almost everyone worked away from home, commuting from one location to the other. We maintained a separate wardrobe for work, even if we didn't have an official uniform to wear. As part of this split personality, we also used different devices, including phones, computers, and PDAs, for work and home.

These days, however, many of us don't need to maintain two phones, two computers, or two PDAs because the services available support both sides of our lives. We take work and personal calls interchangeably, maybe distinguishing them from each other by setting different ring tones or marking certain call types as important. It's convenient and more efficient to integrate our work and personal lives now that e-mail, searches, social networking, online video, and more have become essential parts of both our work and personal lives. And it's increasingly common for companies to let workers combine work and personal lives on a single laptop computer—whether supplied by the company or purchased by the employee.

As a result, many of us feel that work follows us wherever we go, leaving us no hours when we're truly off duty. Others have responded by looking at that permanent digital connectivity as the permission we need to leave the office more often—to go to our kid's sporting match or to work from home.

The growing fuzziness between our work lives and our personal lives also presents a challenge for businesses that market to us. At Yankee Group, that blurring is apparent in many challenges our clients have with planning connectivity products and services. Should a mobile phone subscription where the bill is paid by the consumer, but most of the calls are charged back to her employer, be viewed as a business line or a consumer line? Both, surely: but in designing features, pricing, and marketing plans, how should a mobile operator think about that customer?

The second part of this book looks at the impact of expanding connectivity on consumers, and in Chapter 5 we consider the lives of some consumers in the future, imagining how dramatically the advance of connectivity will change how people care for themselves and their families, how they are entertained, how they participate in government, how they manage their resources.

But that advance of connectivity doesn't stop there. One of the most significant changes brought on by this revolution is how we do our work, and even what exactly it is that we do for a living.

We could debate whether those changes may feel even more profound to us as workers: how much different our work will feel to us in 10 years' time compared to our home lives. But there should be no debate about the potential scope of commercial benefit to businesses from the transformations coming to work from the Anywhere Revolution. Remember the quick cocktail-napkin math we did to guesstimate the scale of connected devices in our future? The proportion of items in the world that facilitate work, rather than consumers' personal lives, is at least one order of magnitude larger, if not two (100 or 1,000 ambient connected items per human compared to 10 personal items). By that measure alone, the profit opportunities for Anywhere leaders will be substantial.

By 2020, connectivity will have changed the workday experiences in just about every commercial and industrial field. It will affect people who work on factory lines, in warehouses, and in hospitals just as much as it will affect white-collar knowledge workers like Sheryl, our advertising industry employee who blended her work and home life so well in Chapter 5. To explore this idea, let's fast-forward again to 2020 and visit a few more of its potential inhabitants, to examine their professions and work environment. Again, as you meet these people, think about your own enterprise and its opportunities for profit. You should be able to spot a number of opportunities to put connectivity to work inside your firm and out. I point out a few of those themes after you meet Ming, Jan, Allen, and Katharine.

Ming Chen, ICU Nurse

For the head nurse of the Singapore intensive care unit (ICU), work is very different in 2020 from the way it was 20 years earlier. Ming is not just the head of the ICU at one hospital. She is the head of the virtual

ICU, which consists of five real ICUs located at five different hospitals around the island of Singapore.

Thanks to the comprehensive monitoring systems assigned to every patient on the five ICUs, Ming and her staff of registered nurses can provide intensive, around-the-clock care without leaving their nursing station. Everything they need to monitor patients' conditions, including up-to-date medical records, dietary information, and stats, are available on-screen to the nurses on duty.

From the virtual nursing station, Ming can set up consultations with the patients' doctors and specialists to determine the best course of treatment. Diagnoses can be made in a matter of minutes, which sometimes makes all the difference. The on-site staff—which still provides in-person treatment—can review the information from the virtual medical station and make the best decisions for the patients.

Rounds are a thing of the past for everyone on the medical team, including the doctors. Connectivity has arrived in the hospital environment to such an extent that the nurses and doctors are always keyed in to their patients' conditions and can be at their bedsides whenever they're needed. The hospital's system monitors everything from the patients' heart rates and oxygen levels to what medications they're taking. If a patient has a problem, the appropriate medical team member is alerted immediately through a wireless transmitter he or she wears; the same device protects the staff members in the psychiatric ward if patients get violent, because they can summon help and be located immediately.

When a patient's treatment involves hospital equipment located elsewhere at the moment—like a portable ultrasound unit or an infusion pump—Ming can instantly locate the closest one that's available, since they're all equipped with wireless connectivity to report on where they are, if they're up-to-date on servicing and maintenance, and whether they're available for use. No more sending an orderly to hunt each floor of the hospital for a missing item only to find it broken and needing attention.

Connectivity helps ensure that patient medications are in stock and correctly cared for until they're needed; with a quick click Ming can explore any of the hundreds of refrigerated storage units around all the ICUs she oversees to see what meds are where, whether they're still viable or out of date, and whether they've been maintained at the right temperature inside the units. As she checks one unit, she briefly remembers the faulty refrigerator in the ward in her very first hospital years ago; you just never knew if the drugs it kept were cold enough on steamy Singapore summer days.

Despite the obvious benefits—chief among them the almost instantaneous access to potentially life-saving information about patient conditions—a lot of people question why nurses and doctors choose to focus their energies on providing this virtual care. After all, it creates an entirely new experience for both the patient and the caregiver; one that, on the surface seems less engaged.

In truth, however, medical professionals like Ming are in a unique position to better care for patients. The monitoring facilitated by the new system offers a much higher level of care to the patient, often in a far less invasive way. Being in a hospital is still a tough thing for many people; making the experience bearable is up to Ming and her colleagues. With the virtual monitoring system, Ming and her colleagues don't have to wake their patients in the middle of the night with every shift change just to check on their condition. Of course, tests are still performed, and there is plenty of interaction between medical staff members and their patients; there are people working as usual, hands on, running tests, dispensing drugs, following up, and discussing care options. Hospital care has become far more streamlined and efficient, however, and everything—from diagnosis to treatment—is done a lot faster and at a much higher level of effectiveness. By working with doctors and nurses at each hospital, who now have far more time to deal with critical patients, Ming can pass on key information about how patients in the ICUs she oversees are responding to treatment. Their care programs are developed from a comprehensive review of information. Tests results are available online anywhere in minutes in most cases, and people receive care and treatment much sooner.

When it comes to treatment, every time a drug is prescribed for a patient, the information is entered into the patient's electronic records system and is instantly synchronized with other medical records systems elsewhere. An alarm sounds if any of the drugs prescribed don't mix well with the other medications the patient is taking.

The patients' families are happy, too, because they don't have to wait for doctors to get an update on their relatives' conditions. When the patients first check in, they're asked who among their friends and families should be notified about which aspects of their care, and as a result, family members can check into conditions if they can pass the online security checks. They can stay up to date without hanging around a nurse's station or trying to track someone down by phone. They don't have to spend time waiting for staff members to hunt down charts and medical records.

Patients can also give temporary authorization to anyone else to log in and see their status from any connected device in the world. A lot of patients use it to get a second opinion without ever leaving the hospital bed. Ming even has one patient with a very rare form of brain cancer who was able to get a second opinion from a renowned oncologist in Mumbai. After reviewing the records, the specialist added the patient to a chemotherapy trial that seems to be working.

Of course, there are still a lot of regulations to ensure patients' privacy—mistakes can be embarrassing for them and expensive for the hospital—but mostly the hospital and the system get it right.

Jan van der Broek, Utility Field Inspector

As a field inspector for a Dutch utility company based in Utrecht, Jan is affected by communications technology far more than he ever expected. He's always worked from his car, but it's his office too these days. Years ago, he used to have to run back and forth to the company's facilities to pick up drawings, blueprints, field notes, and pieces of equipment. Road trips threading through the nationwide traffic jams in the small country used to take up most of his day. But with ubiquitous connectivity, he doesn't even need to go to the office to pick up his tools.

Jan is in charge of marking, inspecting, and protecting the systems owned by his company and buried underground across the region, and he oversees many construction projects. His company often puts cabling in the ground as part of its new development work. Jan is responsible for making sure that the cable location is precisely represented so members of his team can find it when they want to, and so other organizations can avoid it when they need to.

Up until recently, this work involved red-lining: painstakingly measuring the distance of cables from key landmarks and drawing red lines on map to mark their precise locations. Because it is very costly to be even a fraction off the mark, this measuring process, both on paper and in the field, used to take hours.

Whenever people want to put the teeth of a backhoe into the ground, they need to call Jan before any earth is moved. He and his coworkers used to have to go out to each site with blueprints and physically mark the ground with spray paint, doing these markings about 3,000 or 4,000 times a month. This meant that Jan would spend a lot of his time going to the office to get the blueprints he needed for the day.

Every aspect of his work has become a lot easier thanks to a few new gadgets that have emerged. The first and most important item is Jan's netbook, which has global positioning system (GPS) technology built in. The system pinpoints location down to the centimeter when necessary, providing high-resolution location accuracy for online maps. When a new cable is being laid, Jan can do his job simply by standing above the cable. The GPS sends the precise latitude and longitude data over the network to a database where the cable position is captured.

The GPS provides a substantial damage-control element for Jan and his team, as well. If a cable in their system is struck accidentally, the network is pinged and the exact GPS coordinates are determined. Based on the coordinate information, the nearest repair crew is immediately dispatched—all without a single phone call or e-mail. Jan is also put in contact with the team without having to lift a finger. All the necessary information is synched to his hand set.

For the older sites, where the precise latitude and longitude information is not available, Jan still receives an alert message if there's a problem on the system. The only difference is that he needs to download the blueprints of the site to his computer. It's not as easy as the GPS approach, but at least he doesn't have to drive to the office for them, and the service crew is not delayed in getting on-site to do its work.

When it comes to filing reports, things move a lot faster as well. Report requests are made automatically after every service visit, and Jan receives an alert on his computer. The alert includes a partially prepared report for Hans to review and complete. It saves a lot of time and allows Jan to apply his knowledge to make the report information as comprehensive as possible for each service call.

If you ask Jan, it's truly amazing to see access to information become so simple; the time he saves now makes him feel that his contribution to the organization is that much more meaningful. He has more opportunity to exercise his expertise because he's spending less time chasing paperwork and cleaning up mistakes resulting from lack of information. The time saved by Anywhere technology is evident in the sheer amount of work that Jan and his team can manage. They used to handle about 40,000 tickets a year. This year they will get through 60,000 easily, with plenty of time to stop for coffee.

Allen Howard, Commodities Trader

Allen contracts for one of the world's largest trading firms: a Kenyan company that works with hundreds of traders like him all around the world. He used to be a full-time employee, but since the firm extended its benefits, bonuses, and other incentives to its flexible workforce, there's virtually no penalty for not being a captive employee. He much prefers the flexibility of setting his own hours so he can squeeze in as much training as possible for his triathlons. He picks up other work when he wants it, and the result is a nice variety he enjoys.

Allen's focus is coffee, the unit that started everything for this company. He's been working for the company for almost a decade, and he's seen a lot of change. The trading floors were closed years ago, and Allen now works from home.

The biggest part of his job is trading coffee futures. His network operator is able to provide less than one millisecond of **latency** on the fiber optic cable coming directly to his home. This means that there is virtually no delay between the time that he executes a trade on his computer and the moment it is executed online. When the price of coffee futures is changing every fraction of a second, this can mean millions of euros in savings. Even a tenth of a cent fluctuation in the price of coffee can mean millions given the volumes Allen trades. Trading goes on round the clock.

If we joined him in his home office, we'd notice a couple of key differences from earlier trading scenes. The days of trading turrets are gone; in fact, Allen doesn't use computer screens at all. All the information he needs is projected on the walls around him by the onboard projection lamps in his computers. One wall displays meteorological forecasts for the coming year. Another has current trading prices for coffee markets. The other wall is a mash up of all the key applications he uses during the day.

One development is his use of a more evolved version of Twitter, which lets him follow many of his colleagues, competitors, and associates along with real-time commentary from coffee farmers, pest inspectors, and many more professionals up and down the line of coffee production. All the data he uses to make trading decisions are captured in a real-time trading decision database, so he can take time later to review the correlations between his data and his trades. Many of the feeds and information he formerly used he's discontinuing because they were just intermediaries, sometimes with very old data. He finds he makes better trades when he acts on actual farmer crop commentaries and retail store shipment data, because the information is gathered much closer in time and space to the events that affect pricing than anything he had access to in the past. Lately he's been looking for new farmers to follow and trying to gain access to the Carrefour supplier network shipping data.

The other big change is the central role of maps and imagery in his activity. He never thought of himself as a map guy, but rich, interactive, and instantly available maps have become the backdrop of his information consumption in a way he couldn't have pictured 10 years ago. But maybe "map" isn't the right word for what he studies, since he can instantly overlay them with weather, pricing, buyers, historical trends, and more.

Zooming in on real-time satellite imagery of some coffee fields in Venezuela, he notices many plants just starting to bloom. He realizes with a start that a farmer, Vargas, had promised beans from this field to the market much earlier than they're likely to be ready. When he first started out as a trader, his boss told him that the trick was looking for information imbalances between buyers and sellers. Lately with all the real-time information available instantly to anyone anywhere, it seems his success for the company is more about how he puts all the information sources to work to create a useful synthesis of what's actually going on. Anyone could look at that field on the satellite feed; but his talent is anticipating problems by noticing the mismatch between those pretty white flowers and a delivery commitment.

He stopped using e-mail about two years ago, but has no sense of isolation from his colleagues because of all the methods of communications he has—videoconference in particular. Actually, videoconference is something of a misnomer because there are no screens in the office. Instead there are life-size images with quality so good that he can read the notes other traders are taking during meetings. All his communication tools have what he thinks of as "follow-me" capabilities; if he goes to the gym or decides to work on the patio, his telepresence feeds, his mapping configurations, his chat streams with coworkers can appear on any screen he signs onto with his wireless security token.

Physical distances have very little impact on business for Allen and his associates. As his colleagues in other parts of the world sign off for the night and others sign on, his video screens dynamically shift to a view of the new location, whether it's New York, London, Sao Paulo, or Jakarta.

Perhaps the best thing about Allen's work these days is that he can do it virtually anywhere in the world. He has always enjoyed traveling and meeting new people, and he has an extraordinary capacity to do both thanks to new technologies. He doesn't have to leave his home office to meet with people on the other side of the world—not if he doesn't want to. At the same time, though, if he decides to travel to South America or Indonesia, for example, he knows there will be no problem accessing the network, even if he decides to take a trip up the Amazon for a few days. Okay, he doesn't often go this far off-road—he's mostly traveling to check in with his colleagues around the world—but anywhere he goes, his office goes with him. It's totally seamless.

Katharine Goss, High School Teacher

Katharine's been teaching at the local high school for almost five years. Her mother was a teacher, and Katharine was excited to follow in her footsteps. Still, teaching is very different for Katharine. Her mother's stories sound otherworldly, given the sheer amount of change the education system has gone through.

For one thing, Katharine spends a lot more time teaching and a lot less time managing administrative duties. She takes attendance on a mobile device. In the old days, teachers had to sift through page after page on a register, and it was hard to be entirely sure about a student's attendance record without double-checking the register record. Calling the parents to check up on the child was another story altogether. With high school kids, there's always a possibility that the kids are up to no good. There was no way to help the parents keep track of attendance and ensure that parents were aware when their kids were not in school. There are much fewer instances of kids playing hooky these days, since the attendance application automatically calls parents about each student who is absent. An e-mail is also sent to the parents along with a copy of the student's attendance record for the year, just so they're kept informed and problem patterns are quickly and easily spotted—some kids will always hate gym class.

The first period of the day is social studies. The students all open up their school-issued ultra mobiles and go to Chapter 4. Textbooks are a thing of the past, which is great for a whole range of reasons. First of all, the kids have much less to carry around every day, cutting back on the issues of lost books and back problems. Second, the school keeps a stockpile of devices in case kids forget to bring their assigned device. Since all their schoolwork and materials are stored in the network, there's very little personal material on the lost device to have to replace. The content for every subject and level gets updated regularly, with the additional benefit of saving the school a lot of money.

The first half of the session is based on the content from Chapter 4 which focuses on the virtues of the parliamentary system versus that of a republic; the discussion was particularly exciting because the material was updated last night with the news of the coup in Malaysia. The second half of the class session is an online chat with a class from Manchester, England. Today, Katharine's class in the United States leads a discussion about how the electoral college works. Their fellow classmates in the United Kingdom lead a discussion about how the parliament builds coalitions to elect the country's prime minister.

During the next class period, Katharine teaches English to a group of French students in a telepresence class. The students are a small class of 10 students at the sister school in Lyon, while one of its teachers runs a French class for students in Katharine's school district.The system has been in place for a couple of years, and it's amazing how quickly the students in both classrooms pick up on the foreign language. Katharine's American students have excellent French accents, and they're all very enthusiastic to learn contemporary slang to communicate with their French classmates.

The telepresence teaching system is used in a variety of subjects—not just languages. In fact, it's a particularly exciting development because it enables students to take elective classes that the high school doesn't offer, including advanced classes. Limited budgets used to make it difficult for Katharine's school to offer niche courses; it just couldn't support classes that were of interest to only a handful of students, such as Russian, Arabic, the classics, astronomy, philosophy—the list was long. With telepresence, however, the school even has one student taking an archaeology class with a teacher in Egypt.

Although teaching is a vastly different and even more exciting career path than it once was, work is not without its upsets. Katharine and her colleagues face the stress of performance assessments every once in a while—not often, but enough for it to be a little unnerving. All teachers undergo an annual evaluation, and those with identifiable issues, such as poor communication skills or problems managing the new technology, are reviewed more regularly.

During assessments, teachers are monitored from district offices via a live video feed. The assessment includes a thorough review of the teaching material, how the material is being used, and how the students are responding to it.

The nice part about these big-brother assessments is that no teacher is left hanging for weeks on end waiting for feedback. All reviews are posted on the district's social network within 24 hours. That's the site where teachers can interact with other teachers to ask questions, download learning materials, and exchange lesson plans. There's even an area where students can rate their teachers and offer their own slant on the assessment. The transparency can be bracing, but at least Katharine knows where she stands.

Along with her scores, Katharine gets some links to online interactive professional development tailored to the areas in which she needs improvement. She is required to participate, and everything is logged. The sessions themselves are actually pretty useful, with the content adapting to her progress. It's certainly better than the days when they closed the school for three days so all the teachers could sit around while some consultant droned on about techniques that many of them were already familiar with.

One of the biggest changes to the education system is the sheer amount of communication that goes on these days. Schools all around the world are collaborating, sharing resources and ideas. Kids of all ages are benefiting from the opportunity to interact with their peers, other teachers, and top-level professionals. It used to be that only students in the last year or two of high school got the opportunity to hear from experts in different fields, attend university lectures, enjoy museums, and have access to all sorts of other dynamic learning resources. Many students in Katharine's class already spend their free time following real-time feeds from universities they're interested in, and many times this means they are reading up on important developments in scientific and professional fields at the same time that the university's full-time students are doing the same thing.

These days, even kids in kindergarten are being taught how to explore the world and absorb information. As a result, students are more focused and more enthusiastic about learning because it's no longer about core curricula and SAT scores. Everyone has to learn the basics—the core subjects are still considered as important as ever—but there is much greater opportunity for students to pursue their own interests and try new things within the education system. Even the less academically inclined have the opportunity to find their footing and get a sense of where their strengths lie. The result is a far more collaborative, productive, and nurturing environment for all. In real terms, this has translated to higher attendance rates, lower instances of behavioral problems, and excellent results for standardized testing.

Becoming Anywhere Workers

These vignettes give a sense of the broad set of profit opportunities that lie ahead for enterprises from the introduction of more connectivity in workers' lives, either by deploying them inside your own organization or in developing those solutions for others to deploy. The key themes I see in the transition of these common roles into Anywhere functions involve changes in tools, the nature and immediacy of information, the significance of location, and the very nature of the work they do. Let's look at each of these changes. (See Figure 7.1.)

First, **Anywhere Workers** in 2020 will have a whole host of new tools to help them do their jobs. Generally, people will be using fewer tools, but those tools will be more general-purpose. Think about the nurse's system for monitoring patients at five different hospitals, or the

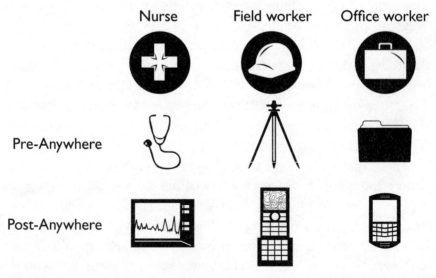

Source: Yankee Group

Figure 7.1 *The tools of Anywhere Workers.*

assigned mobile device given to students to replace perhaps five or six textbooks. Think about the single device that enables the field inspector to map the exact coordinates of his company's vast network of underground water systems. Anywhere Workers of the future will have fewer devices performing more sophisticated tasks. The steadily falling price of connectivity and computing means that tools will become smarter and will be networked more effectively. Which of those tools are ones that your firm might develop and offer? Which of those tools, if rolled out in your organization, could contribute to reduced labor expenses and thus a better bottom line?

Workers will also be dealing with more "bits" than "atoms." After years of talking about the paperless office, the advent of pervasive, capacious networks combined with more suitable electronic reader devices will finally mean that Anywhere Workers will consume fewer trees, helping ease a major environmental concern. Imagine a world in which everything from student textbooks to corporate handouts and client publications do not have to be printed en masse and then reprinted a few years later when new, more up-to-date information becomes available.

The nature of information will change, too, and we will process it differently. Video and other forms of rich multimedia will be not the spice, but the daily norm of how many people will work. We can only wonder at the impact on interpersonal communications when professionals can see each other in real time no matter where they are in the world. While we're part of the way there now thanks to systems like Skype, which has free video calling, Anywhere Workers will have video of an entirely different caliber. Images will be lifelike and immediate. Richer collaboration tools will be an explosive part of the connectivity revolution in the workplace, one we explore in the next chapter. How could your firm work more efficiently with better real-time, high-value connections among workers, partners, and customers—speeding time to resolving customer concerns, improving on the quality of output through greater teamwork?

The immediacy of information will also be another major change. In our vignettes, information is captured right at the place and time of the activity. When the teacher takes attendance the messages to parents are sent out within seconds. The ICU nurse has up-to-the-second information about all her patients as well, and the information is delivered right at the point of need. In all our scenes, Anywhere technology has dramatically reduced or even eliminated time delays.

Anywhere will change the significance of location for employees in all types of industries. Many of them will not be required to be in one specific physical location—the nurse and the trader are able to maintain a virtual presence rather than a real one in order to do their jobs. Workers whose jobs do require them to be on the move will find it easier and more productive; our utility inspector wastes much less time going back to the office to collect physical information. Where are the cost savings in your enterprise from reducing or eliminating the need for a physical presence everywhere you have workers and customers, or from improving turnaround faster than your competition by leveraging a 24/7 workforce positioned around the world?

There are changes to the *nature* of work, too—the actual tasks that an employee does. In all four vignettes, the workers are doing different things in their jobs from what people in those roles do today.

One of the reasons that work itself will become so different is that larger tasks will be broken down differently inside organizations. The changes the nurse experiences also affect her coworkers in physical plant maintenance, and the changes the trader describes imply different ways of allocating activity. Their employers assign responsibilities to these employees in different ways from the ways hospitals or financial services firms do today.

This creation of a more united, aligned work system in our Anywhere future will be particularly important. In *Grown Up Digital*, author and social scientist Don Tapscott explains the Net Gen, as he calls the echo of the baby boom, is more demanding of employers than prior generations. "[Net Geners] come into the workforce with very different attitudes about collaboration, about having fun at work, and about being free to work when and where they want. I think we're seeing the early signs of a major collision between the freewheeling Net Generation and traditional boomer employers. It's a clash between two ideas of how work should work."

As LiveOps chairman and CEO Maynard Webb points out, "Work today is broken. One out of four people have been in their jobs less than a year; one of every two, less than five years. Companies are trying to find people, and people are trying to make trade-offs between having a life and having a career." He points out that more dynamic workers, many of them freelancers, are already beginning to build careers around their competencies while working for many different companies on their own terms. And it's not likely to change back; the U.S. Department of Labor predicts that today's students will likely have had 10 to 14 different jobs by the time they are 38. Connectivity-centric technologies including telepresence, smart grids, collaboration tools, and much more can help overcome the loss of continuity and context that can come with a less permanent, less stable workforce by making information capture automatic and the allocation of work to transient workers easier to assign and manage.

With Anywhere technology in the health-care system, a new nurse starting work will be able to come up the learning curve faster and to stay on top of his or her responsibilities better. Virtual wards will provide workers with a wealth of information instantaneously. New nurses won't have to worry about learning which doctors are treating each of the patients; if there are any emergencies or changes to a patient's condition, the appropriate doctors and medical team professionals are notified automatically. The focus of the nurse's job is to treat patients. Similarly, teachers and utility workers are saved a lot of time stepping into new situations. Student records and class schedules are all accessed via the network; utility workers like Jan don't have to hunt for paperwork before their next service check; commodities traders don't have to deal with filtered information but instead receive realtime updates from sources relevant to making trading decisions.

By 2020, more and more workers will spend less and less time in traditional offices. Many will enjoy the advantage of being able to work from home or any location that best suits their needs. More people will find themselves working as free agents, serving multiple companies and employing multiple types of connectivity devices to do it, many of them the consumer tools they are already using at home. Anywhere will move many of us toward being virtual employees. Will the Anywhere Enterprise of the future be a more virtual one? We explore this in the next chapter.

The way people work in 2020 will be profoundly different as work life becomes more thoroughly integrated and intermingled with personal life. We will be able to work from Anywhere, on the go, at home, at the office, even if the companies we work for have offices in central locations. If they do, the office will be more a choice than a necessity, because executives and managers will be able to monitor the work of their employees from any location.

Our use of devices will be different as well, and Maynard Webb points out the implications for employers: "Enterprises have to understand that people will want to use their own handheld devices, cell phones, or whatever, and your architecture should enable that to happen. You'll have to dial up security in a big way, so that regardless of where they are, they can get what they need from the company's networked assets."

So often when people think of technology, they concentrate on the image of a knowledge worker sitting behind a desk and staring at a computer screen all day. They're thinking chronic carpal tunnel syndrome. In reality, office workers stand to gain a lot from Anywhere advancements, but they are by no means the only group of professionals who will benefit. Everyone stands to gain, and blue-collar workers may well gain more. After all, it's the construction worker and the farmer who have been starved for information because of device constraints like price, weight, fragility, and poor quality network access. As technology becomes cheaper and more sophisticated and as information is pushed to these workers by the Anywhere Revolution, they can start to respond in kind by sending information back. The result is the ability to make better decisions faster and more effectively.

Now that we have a picture of how people the world over will change their activities both as consumers and as workers as a result of Anywhere, let's pull back and look at how commercial enterprises will be changing in more systemic ways. These larger transformations will provide yet another rich lode of profit opportunities in the development and deployment of ubiquitous connectivity.

Chapter 8

The Enterprise Goes ANYWHERE

"With connectivity as your platform, you can have a boundaryless enterprise."

—Doug Hauger, general manager, Business Strategy, Windows Azure, Microsoft

What happened on October 27, 1986?

Ask that question in Boston, where I live, and if you get something other than a blank stare, the most common reaction will be a moan. That's because it was the day that the Boston Red Sox baseball team lost to the New York Mets in the final game of the 1986 World Series. But despite the tragedy of that loss for Bostonians, even we might admit that another event on that day had a more profound effect on the world at large. In fact, that date is so significant that it has a name anyone in the financial industry recognizes: Big Bang.

Big Bang was the start of the deregulation of the London Stock Exchange (LSE); the beginning of a total transformation in stock trading, moving price information from regulated brokers shouting on a London trading floor to an electronic quotation system. That transformation was completed 10 years later when the LSE replaced that first-generation system with a fully automated and networked trading system that eliminated the need for brokers to call in their orders. Within a decade, the London Stock Exchange had been transformed from a *place* to a *network*.

This change might sound important only to financial services executives and technology professionals, but its effect on the economy has been profound. According to data from the London Stock Exchange and the Bank of England, at the time of the 1986 Big Bang the shares traded on the London Stock were worth roughly £161 billion. By the twentieth anniversary of Big Bang in 2006, issues traded on the LSE were worth more than £2,496 billion. In 1986, no non-British issues were traded on the London Exchange; by 2006, foreign equities numbered more than 2,700. By moving its operations from the constraints of the physical world to a network, the London Stock Exchange changed London from a lagging player in global finance to the number two financial market globally after New York (and that ranking is debatable).

So what does this all have to do with Anywhere?

The Anywhere Network will similarly transform businesses into what we'll call Anywhere Enterprises (more on that in a moment). In the process, the analysts at Yankee Group predict that the global revenues associated with the **information and communication tech-**

nologies (ICT) powering these transformed businesses will explode from today's estimated $2.2 trillion worldwide to over $4 trillion by 2016. And just as in London with the Exchange, the increased reach and value of businesses that mobilize and capitalize on the network will be many times greater than that figure.

That's a big claim. After all, in 2008, the International Monetary Fund estimated that the entire gross domestic product of all the countries in the world was only $61 trillion. Can the Anywhere Network really add a sizable fraction of that amount to our global economy in just a few years?

It will. And the business and technical underpinnings of Big Bang provide a useful road map to understanding why.

Big Bang changed three major factors in the London Stock Exchange's operations: electronic trading floors, remote workers, and more flexible jobs at the Exchange. Yet just those three changes dramatically changed the path of that hallowed institution. (See Figure 8.1.)

The ubiquitous connectivity emerging with the rise of the Anywhere Network offers similar opportunities for business transformation. Three parallel factors will create what we term the *Anywhere Enterprise*, which we define as an organization that uses the Anywhere Network to radically transform what it does and how it does it.

	Before Big Bang	*After Big Bang*
Trading floor	Physical	Electronic
Workers	Local	Remote
Nature of work	Specialized	Generalized
Value of traded shares	£161 billion (1986)	£2,496 billion (2006)

Figure 8.1 *Big Bang reinvented the London Stock Exchange.*

	Pre-Anywhere Enterprise	Post-Anywhere Enterprise	Primary benefits
Anywhere IT	Physical	Virtual	Cost savings, operational flexibility, better use of capital
Anywhere Workers	Local	Global	Larger talent pools, more employee satisfaction, lower real estate costs
Anywhere work	Tied to business location	Untethered	More customer satisfaction, more agility, more fluidity, more rapid change

Figure 8.2 *The evolution of the Anywhere Enterprise.*

In the next three chapters, we describe the details of these three transforming factors. Each of these provides profit opportunities—either in the form of cost savings or greater revenues. (See Figure 8.2.)

Infrastructure: Virtual Services Replace Physical Technology Assets

The first and most obvious change the London Stock Exchange embraced was moving to an electronic exchange from a physical one. The LSE called its first system that allowed brokers to get prices electronically the Stock Exchange Automated Quotation system, or SEAQ. That first system was replaced 10 years later with the Stock Exchange Electronic Trading (SEET) system, which not only provided quotations, but also allowed direct electronic trades. Both were important because they allowed LSE members to do all the things they had done for more than 200 years, but without the physical trading floor. Said another way, they replaced the physical trading floor with a virtual one.

This idea of **virtualization** makes sense for a stock exchange, but how does it help a more traditional company that makes cars or provides accounting services? Are they going to virtualize physical assets like assembly lines or offices?

Obviously not. But in the developed world, there is a business area nearly every company owns that can be easily virtualized: the corporate data center. More specifically, we can virtualize the servers, software, and other assets related to providing a company's technology platform.

Servers are the hardware platforms for corporate software applications like transaction processing, customer relationship management, and e-mail. Most companies consolidate these servers in tall racks housed in dedicated data centers around the world. These centers have been created because of the rapid growth in these business applications over the past two decades. Companies did this because servers based on personal computer technology were cheap on a per-unit basis, and demands for those applications continually increased. The result: Corporate data centers are experiencing "server sprawl," now housing thousands of PC-based servers representing billions of dollars of capital investments. Chip maker Intel runs more than 100,000 servers for its chip design and simulation work alone.

While this sounds simply like the cost of doing business, the vast majority of computing servers that run those applications sit idle or unused most of the time. According to the Uptime Institute, 80 percent of data center servers that run corporate applications have average utilizations of less than 30 percent; even worse, up to 30 percent of those servers are completely unused, having been provisioned for an application that the company no longer runs and thus becoming a stranded asset—one that in many cases still consumes power. When we compare those figures to those from the 1970s and 1980s, when servers were used 80 percent of the time or more, it's clear that enterprises have opportunities to make better use of these capital-intensive corporate assets.

Server virtualization addresses this problem. It allows business managers to address the server sprawl that infects today's corporate data centers. Virtualization runs many applications and their associated operating system software on a single server computer while allowing data center managers to operate those applications just as they would on dedicated servers. By putting multiple applications and operating systems on a single server, virtualization drives up the utilization rates and gives hardware a workout, squeezing as many processor cycles from the hardware as possible. This also makes the entire concept of a server more fluid and flexible. Because these virtual servers are really just computer files, entire workloads can be backed up, copied, or moved across the country in seconds.

This model of virtualization has become wildly successful in businesses, governments, and service provider networks where thousands of servers typically run business applications. According to a study done by IBM, the savings from virtualizing servers in a typical data center can be as high as 40 to 50 percent, with an average data center savings of $1.4 million per year. Further, as we discuss later in this chapter, reducing the number of physical servers also reduces the need to house and cool them, generating additional savings in both dollars and carbon emissions.

But server virtualization is only the first step in moving a business from physical to virtual infrastructure. The second step is to stop owning the servers altogether.

If servers are just files that can be moved over the Anywhere Network, these resources could live anywhere, not just in the corporate data center that the business owns. This idea—that servers can be rented from the network "cloud" rather than owned in a corporate data center—has given rise to an entire industry devoted to the broader idea of moving all applications and services out of the company's data center and into the network itself. The concept is called **cloud computing**.

One of the companies that has popularized cloud computing services is the online retailer Amazon, which first started delivering its Elastic Computing Cloud (often abbreviated EC2) services in 2006. Designed originally as a way to make use of Amazon's excess server capacity during nonholiday periods, it's now become a widely used service by corporations both large and small.

The reason lies in the opportunity to increase agility. Large corporate IT operations struggle to be agile and responsive to their customers. But complex purchasing processes, diverse responsibilities, frequent interruptions, and plain old not-invented-here thinking bedevil that ambition. The result is that IT centers can take weeks, months, and even years to respond to an organization's need for more computing capacity for new projects or for sudden surges in volume. When the data center does respond but then the need abates, a second problem emerges—that of having used the company's funds to acquire technology that may no longer be needed.

The ideal way to deploy IT resources would be to acquire them at the snap of a finger and to release them just as quickly, moving from a fixed expense to a variable one. "Drug maker Eli Lilly had an in-house computing environment that was very efficient, but not agile," explains Amazon CTO Werner Vogels. "To get a single server set up for a new project required six to seven weeks. With Amazon's EC2 service, it took them five minutes to request the capacity and get it." And when they're done with the temporary capacity, it takes only a minute to release the resources and stop being charged for them.

Cloud-based services allow businesses to do things that weren't previously possible in the same time frame and budget. Virtualization takes an entire workload and allows it to run anywhere. Cloud computing provides the ability to rent space on the network where those virtualized workloads can run on demand, as opposed to purchasing the space and being stuck with it whether you need it or not. The combination of these two techniques, along with the maturation of collaboration tools to better unite the efforts of knowledge workers no longer sitting at adjacent desks, is what we call **Anywhere IT**. (See Figure 8.3.)

Just as with the London Stock Exchange, Anywhere IT removes the cost to an enterprise of acquiring and operating vast amounts of commoditized physical infrastructure for computing: servers, storage, and more, and replaces it with lower-cost resources located elsewhere and available on demand. More importantly, these virtual computing assets are more flexible and agile, able to respond to business needs more quickly without large amounts of capital, labor, and real estate. But there's more to this virtualization story than lower costs. The Anywhere Network also changes an organization's reach, as we explore next.

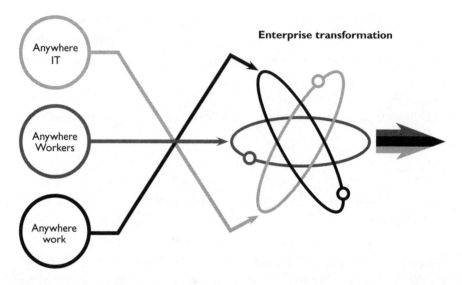

Figure 8.3 *The Anywhere Enterprise evolves from the fusion of three elements.*

Workers: Anywhere Workers Don't Have to Be in the Office

By turning the physical trading floor into a virtual one, the London Stock Exchange created a new opportunity for itself and its members: the ability for brokers to execute trades without being physically at the LSE. Combined with changes in ownership rules, this change allowed the LSE to become a global exchange nearly overnight. In 1985, the London Stock Exchange had zero international members. As of 2009, the LSE had more than 3,000 firms from 70 countries allowed to trade on it, making it the most international of all exchanges.

This ability of the LSE to engage a global market of financial services firms is a good metaphor for what happens when organizations virtualize their IT infrastructures. Traditionally businesses have kept their business applications inside their physical walls, protected by security firewalls, that, as the term suggests, limit outside access to those applications. Like the LSE, enterprises that can remove the barriers to remote participation should gain new leverage from human capital and the ability to assemble top-notch teams anywhere in the world.

Anywhere Enterprises can profit first from this transformation by better accommodating the mobility of their employees today. Businesses are at a tipping point in the evolution of the workplace that will transform how and where people work. In 2004, Yankee Group's enterprise surveys showed that 30 percent of business employees were mobile (which we define as spending more than 20 percent of their time away from their primary work location). By 2009, that number had risen to 43 percent, and this trend shows no signs of stopping.

Anywhere working—the freedom to work from home, a café, the commuter train, a hotel room, or anywhere in between—is becoming a prerequisite for employees. Our surveys of workers in North America show that nearly three-quarters of them believe that allowing employees to work from home benefits the company. We also ask employees to identify the single most important thing that their company could do for them to improve their productivity. Even with a wide array of answers to choose from, such as relaxing security policies, choosing their own corporate devices, or adding new technology, the top response is the ability to work from home. Enterprises that fail to respond to this trend will be at a severe competitive disadvantage in the war for talent.

This trend of remote working, though, requires a shift away from the traditional models of hard-wired business software applications that fetch corporate data from across a **virtual private network (VPN)** or remote connection. Picture a typical corporate executive working today from her laptop in a hotel room. She has trouble connecting to her office wireless network, so she defaults to the guest network and sets up a VPN connection. She then launches several stand-alone applications—let's say the company's customer relationship management application, its order entry system, and an inventory management tool, toggling manually among them to figure out if her customer's order has yet been processed and if there is enough product in stock to ship it today. One application stops working, so she struggles to reach her company's technical staff to troubleshoot the problem. The corporate infrastructure that supports her costs millions of dollars, and it takes her one to two hours of her workday to try to answer the customer's simple question: "When will I get it?"

Contrast that typical experience with a young child sitting at a video game console in his bedroom. He picks up a headset, plugs it into his game controller, and can instantly chat with a community of gamers. His game controller and the console connect over wireless networks, and his voice is sent over the network using **voice over Internet protocol (VoIP)** technology—yet he doesn't think about the supporting connectivity. The technology just works, it costs under $1,000, and it is far more effective than that used by our frustrated executive. The forces of the consumer marketplace—huge scale accompanied by a terrifying demand from buyers for simplicity—have brought the child a better computing environment than the business executive has. What if your enterprise employees could be that effective away from the office?

Today's Anywhere Worker trends are driven not by enterprise models of technology adoption but by the trends we experience in consumer electronics—a "consumerization" of enterprise computing, if you will. In prior chapters, we saw how technology is becoming more personalized and portable and how content will migrate from our computers, cell phones, iPods, and games consoles onto the network that connects all these devices. The consumer netbook—a low-cost, network-enabled laptop with minimal onboard storage—provides us with a better vision of what Anywhere Enterprise applications for remote workers will be like over time. With netbooks, you can access everything over the Internet. You don't need to store software or personal files on your computer anymore because all these things can be kept more efficiently and more safely online.

Just as servers are becoming virtual servers, the corporate applications on a worker's laptop, commonly referred to as her computer "desktop" or "image," are also being virtualized. The benefits of a worker being able to access a virtual corporate desktop computing environment from any computer are extraordinary. At one end of the spectrum, they limit the security risk from loss or theft of computers with sensitive corporate data on them. They also simplify laptop provisioning for workers by letting them use their own laptops and providing just the security and authentication services needed for them to reliably identify themselves to gain access to those assets from whatever device they happen to have access to.

How does this help our stranded sales exec trying to tell a customer if his order is moving through her company's systems? Take a simple-to-use netbook, a security token that authorizes the executive to use a public network rather than the VPN, add the right network-based software to integrate data from those siloed applications with onboard chat and other self-service tools—and her ability to work smoothly while at a remote location begins to look more like the young gamer's experience.

Maynard Webb of LiveOps describes the trend this way: "If you look back 15 years ago, the computing power inside a company was ten times what you had anywhere in the general consumer world. But the consumer computing world is now three times as powerful as the business world in the amount of information you can get at your fingertips. Young employees are used to the consumer model, and enterprises have to be nimble to live up to their expectations." Furthermore, from this latest generation's perspective, a lot is changing. They feel increasing pressure from rising gas prices, hassles in driving to work, and a fixed number of hours in the day. "All of these trends are accelerating. Young people are looking for alternative lifestyles as employees."

Bolstering this trend of Anywhere Workers is the fact that some governments have created incentives and mandates for corporations to allow remote working. To alleviate wretched traffic congestion, reduce carbon emissions, and confront a forthcoming labor shortage, the Netherlands created tax incentives for organizations that enable remote working. However, in the event that dangling the carrot of tax incentives to Dutch organizations isn't enough, the government also wielded a stick that would further compel organizations to transform. Dutch labor unions and the federal labor board worked with some of the country's largest employers to set remote working mandates with mandatory goals that must be met beginning in 2009.

But perhaps the final nail in the coffin of trying to keep workers anchored to an enterprise's physical plant is that Anywhere Workers allow businesses to realize huge cost savings—because it eliminates the need for the real estate to house them. IBM saves $110 million per year because more than one-third of its 300,000-plus employees work at home. Additionally, remote working can lead to a more satisfied

workforce with lower turnover rates that in turn reduce replacement and recruiting costs.

If you're already working in an IT capacity in your enterprise, you're probably well aware of these technology changes and many more related to the expansion of the network. (See Figure 8.4.) If you don't, one of your first steps to help your business profit from the Anywhere Revolution should be to develop at least a passing fluency with the core technologies, recognizing how they help collapse cost in the firm. From software-as-a-service (SaaS) to enterprise mobility, these innovations all depend on the Anywhere Network to bring information and assets closer to the worker and reduce complexity in their delivery.

Anywhere IT is a commitment by an enterprise's technology leadership to make critical software applications cheap and easy to deploy on the network so that businesses can scale up and down in response to demand and so that workers can access information assets from wherever they are. The workers of the Anywhere Enterprise will have higher job satisfaction while helping the firm reduce costs. The right tools and policies for Anywhere Workers will provide access to a global workforce. That leaves one more component of the Anywhere Enterprise to examine: the transformation of work itself.

Work Itself: Collapsing Time and Distance between People and Things

In 2006, the *London Telegraph* ran a retrospective looking back at the 20 years since Big Bang. In it, Michael Marks, one of the proponents of the change, talked about its impact. Even after having sold his LSE jobber firm Smith Brothers to Merrill Lynch because it wasn't big enough to compete, he noted that, "Big Bang turned us from the nicest club in the world, a great men's club with good jokes, nice lunches and wonderful people, into a business. It may be greyer, more aggravating, more regulated and all the stuff we moan about, but London became more professional, and liquidity on stock exchanges and international markets gravitated there."

Unified communications:
Definition: Integration of non-real-time (e.g., e-mail, voice mail) and real-time (e.g., voice, video) communications
Anywhere benefit: Faster customer response times

Wiki:
Definition: Shared electronic repository for collective corporate information that anyone can update and edit
Anywhere benefit: Faster innovation cycles through improved collaboration

Enterprise social network:
Definition: Online community of people who share interests, activities, or expertise
Anywhere benefit: Capture and reuse stranded institutional knowledge

Mobile VPN:
Definition: Continuous access to network resources and information on mobile devices
Anywhere benefit: Improved field worker productivity

Instant messaging:
Definition: Real-time communication between two or more people based on typed text
Anywhere benefit: Faster answers to customer questions

Mobile e-mail:
Definition: Reading, replying to, and forwarding e-mails using mobile devices
Anywhere benefit: Shorter sales cycle

E-Records:
Definition: The digitization of paper-based records and forms
Anywhere benefit: Cost savings through remote working enablement

Field force applications:
Definition: Applications that streamline processes for field force workers via technology
Anywhere benefit: More work order completions per week

Desktop virtualization:
Definition: Hosted and centrally managed desktop that gives end users a full desktop experience
Anywhere benefit: Reduced desktop maintenance and support requirements

Cloud-based services:
Definition: Accessing corporate infrastructure or applications over the Internet
Anywhere benefit: Reduced capital expenditures

Desktop video:
Definition: Ability to send and receive video from corporate desktops
Anywhere benefit: Reduced sense of isolation among remote employees

Machine-to-machine communication:
Definition: Networked sensors that automatically collect data for reporting and analysis
Anywhere benefit: Reduced energy costs due to remote monitoring and control of temperatures

RFID (radio-frequency identification):
Definition: Tracking of goods using a connected tag affixed to the assets
Anywhere benefit: Reduced inventory costs due to fewer lost assets

Source: Yankee Group

Figure 8.4 *Some Anywhere Enterprise technologies.*

I'd say that more simply: When your enterprise can both draw on the assets of the entire world and compete with it at the same time, the way you work changes.

In an Anywhere Enterprise, the nature of the worker and work itself change. Work no longer is a place and an activity; it becomes merely an activity. Electronic business applications and information move outside the enterprise firewall to the cloud, and workers essentially can do the same. If the "what" (platform) and "where" (location) of IT become irrelevant, the same goes for end users.

Doug Hauger, who leads Azure, Microsoft's cloud computing initiative, sees not just the potential for new forms of relationships among workers, but also among business partners. "Autodesk, the computer-aided design software firm, gets this right. Their newest product direction enables collaboration between the many different parties who design and construct buildings and infrastructure—the architect, engineer, bill of materials, all the way to the guy delivering the cement on-site. They're focused on bringing what's inside the walls of one enterprise and extending it out ubiquitously to all the project's partners."

As we discuss in Chapter 7, many people will be going to work without actually going anywhere. Nurses will be able to monitor multiple wards at any given time from a centralized location. But those wards won't just be one of several floors, as is common today. Nurses will be able to monitor wards that are located miles apart, collaborating with other health-care professionals all over the world to provide optimum care to patients in the most cost-effective and efficient manner possible, largely thanks to the Anywhere Enterprise and the cloud system that defines it.

But the transformation begun by how workers do things using the Anywhere Network is just the start. What becomes more transformational is when we can connect better not only to other people but to *things*.

Medical facilities like SUNY Upstate Medical University in Syracuse, New York, are attaching active wireless radio tags to medical equipment; with Wi-Fi access points serving as readers, infusion pumps and other mobile treatment devices can be located instantly. Terry Wagner, the facility's CIO, says, "Lots of equipment moves around; some things get lost, others have to be sent for sterilization after they're used." If 200

searches per day—each of which might require 15 minutes of a nurse or engineer's time on average—are eliminated, a hospital like hers can redeploy almost $1 million per year in labor to patient care. If higher utilization also reduces the number of those devices required to service the hospital's patients, then there are direct savings realized from lowering purchase or rental expense too.

Viracon, a specialty glass manufacturer based in Owatonna, Minnesota, builds customized glass panels for skyscrapers. Because every piece of glass it builds is custom manufactured, tracking those unique glass products as they pass through manufacturing facilities measuring millions of square feet used to be a logistical nightmare. Now, Viracon tags every piece of glass it makes with a RFID tag and monitors its progress through the facility using a wireless network installed for that purpose. Because the firm is able to better track its products through its facilities, it has reduced scrapped product by 65 percent, paid back its investments in the connectivity technology in less than six months, and reduced the amount of inventory it holds, thereby increasing the business efficiency of the company.

When people can find objects electronically instead of physically going out to look for them, business performance improves. But the opportunities go beyond just asset tracking. "Anywhere you have parts, tools, or people as part of your workflow, you can ask how to improve the process," says Antti Korhonen, CEO of real-time location systems innovator Ekahau. "But you have to make a process visible before you can see its failings." By tagging people and assets with low-cost wireless transmitters using the in-house Wi-Fi networks that more and more companies have already, the process that is actually used—versus the process as it's written down somewhere—comes alive. Manufacturing operations are starting to realize the tremendous insight embedded in this initiative. For another example of an Anywhere Enterprise adding connectivity to its physical assets to save time and money, see the sidebar on pages 158–159, profiling Hewlett-Packard's deployment of RFID technology in its printer manufacturing and distribution operations.

Russ McGuire, vice president for corporate strategy at Sprint/Nextel, tells how a similar move—using connectivity to put information at the point of need—changed car rentals. "At least a dozen years ago, Avis did

Becoming an Anywhere Enterprise: Hewlett-Packard's Move to RFID

Why does an enterprise decide to use connectivity to collapse the costs hiding in time and distance? Sometimes, it starts from external pressures from a business partner.

Christian Verstraete, chief technologist for Manufacturing and Distribution Industries with Hewlett-Packard, describes the global technology firm's foray into the addition of connectivity to its printer manufacturing and distribution process:

Q. *What made H-P add RFID technology to your manufacturing?*

A. In 2006, we got a now-famous letter from Walmart saying, "You're one of our top suppliers, so please put RFID on your shipments to us." They were automating receipt of products at distribution centers.

But it was the right thing for H-P to do. Large manufacturing companies, particularly electronics operations like H-P, are merging and consolidating to gain more scale. And we are outsourcing every noncore activity in the enterprise. So our supply chains are increasingly complex. At H-P, we have 400 to 500 companies all over the world who are part of our manufacturing process. Every year we make 55 to 60 million printers, and nearly 50 million PCs. Every second of every day we buy $100 worth of memory alone. We need to know where our stuff is at any moment.

At the same time, the shelf life in a store of a new PC model is typically six months. Amidst increasingly complex supply chains, intense price pressures, and short shelf lives, we have to continue to make money.

Q. *What did you do as a result of Walmart's requirement?*

A. By 2006, passive RFID tags—the stickers placed on items that are "read" by an RFID reader to report on what they are—had dropped in cost to about 15–20 cents each, making it fairly inexpensive to do a "slap and ship" RFID application. That's when you put a tag reader at the final packaging station before it leaves the factory, stick a tag on the pallet, and ship it so Walmart could have what it wants. But we decided not to do that and to start tagging product elements before manufacture was completed. We equipped the entire manufacturing line with RFID readers.

Q. *What were the key benefits from that decision?*

A. We looked more closely than we ever had before at *time*. With RFID readers at each stop telling us when a chassis really arrived and left, we suddenly knew the actual time it took us to move each chassis through every step in the process.

Take for instance a step that takes an average of 12 minutes to complete, but has a deviation across multiple observations of plus or minus 3 minutes. To calculate the production capacity of a manufacturing line, we have to use the worst case for that step; in our example, it took 15 minutes.

So we asked why. Can we reduce that deviation? If we can reduce the variation in process time to plus or minus 1 minute, then the total is now 13 minutes, not 15. We looked at step after step in the production line with this actual timing data and found excessive deviations that we were able to reduce to improve throughput.

And we reduced a number of steps. For instance, at the point we load a pallet of printers at a shipping dock, we used to have people holding manual bar code readers to read the serial numbers off each of the printers; that's how our system knew which actual printers got shipped where. Once we went to RFID tagging, we just put the product on the pallet; at the moment the pallet gets wrapped up, an RFID reader at the shrink-wrap station automatically logs which printers are going out. That is time won through fewer steps.

Q. *What kind of reaction did you have from employees?*

A. We did have to do lots of worker education. One of the worries in the factories was, "Now you're going to control me." Another reaction was, "I want to make sure my union is involved, to see what they say." We had to educate union leadership, explain what we were doing, and why we were doing it. It took about six months for people on the manufacturing floor to get comfortable with the changes.

Q. *What was the return on investment for H-P?*

A. RFID technology has paid us back, without question. And that's even with paying a higher price to do it than companies that have come to this later, since RFID prices are dropping quickly. Even though we paid more because we started earlier, we got a payback on it in about six months.

The implementation of RFID reduced the amount of our work in progress by 18 to 20 percent. It doesn't affect revenues as much, because adding RFID doesn't add value to the product itself. But it has helped us with some of the insight we get from Walmart, we see their inventory, where throughput goes, help them deliver to the right places.

And Walmart's happy too; according to a study from the University of Arkansas, Walmart has had a reduction in stock-outs (shortages in product) of 11 to 14 percent from its key suppliers resulting from their implementation of RFID.

something that all of us can connect with. The old way was, you parked the car, pulled out the folder, wrote down the mileage, stood in line, then they keyed it in and told you how much to pay. Hopefully you caught your flight. But when they built connectivity into that process, they put the check-in workers where I wanted them to be as a customer, which was wherever I pulled up in the car. Without even having to find a pen and deal with the folder, suddenly I was getting on the shuttle. Did that reduce costs for Avis? Maybe. But it changed the rules of the industry by creating so much value that everyone else had to follow suit."

It's amazing what you can do when you free essential business functions from the confines of the office.

The Anywhere Enterprise Creates a New Competitive Landscape

To this point, we've described an Anywhere Enterprise as one that uses virtual, cloud-based services, employs workers who can work from anywhere, and reinvents its processes to collapse the costs inherent for both the firm and its customers in extended time and distance. But that dry description is a bit like describing sashimi as slices of cold, raw fish; it doesn't do justice to the product. Worse, it suggests that these changes will be benign and easy to adopt at whatever tempo businesses may think is convenient.

Nothing could be further from the truth. Why? Because as entire countries move past the Anywhere Tipping Point, businesses will inexorably transform with them. Said another way, even if your business doesn't become an Anywhere Enterprise, you will encounter Anywhere Enterprises as your competitors. Those competitors will have some impressive qualities you'll have to deal with: they'll be more efficient, greener, and very different in what they actually do. Let's look more closely at each of these assertions to understand why.

Anywhere Enterprises Will Be More Efficient

In an Anywhere Enterprise, essential business functions move from requiring capital expenses—up-front investments of cash—to operational expenses, part of the day-to-day costs of running a business that are usually paid for from the cash flow from operations. This change has precedent: Manufacturers at the turn of the twentieth century stopped buying their own electricity-generating equipment when they could buy electricity from power distribution companies. Their generator capital expense (one-time big investments, "capex" for short, called "below the line" in accounting speak, requiring big bites out of reserve cash or worse, complex financing programs) was replaced with power operating expense (on-going costs, also called "opex," typically smaller, appearing in an annual budget). In the last 20 years, airlines have stopped owning expensive aircraft, preferring instead to turn the capital expenses used for purchasing into operational ones through leasing, conserving cash. The same will happen with Anywhere IT; technologies such as virtualization and cloud computing mean that businesses will be able to rent their business applications instead of buy them. Capex-intensive hardware and software will become cheap opex-based services.

Some figures can help provide a gauge of how much more efficient this approach will make businesses. According to the U.S. Department of Commerce's digital economy report in 2003, the United States spends roughly $300 billion each year on information technology equipment and software. If enterprises were to rent access to virtual versions of those same applications, they could do so for roughly 10 percent of that price or $30 billion. The result: in the United States alone, businesses would save more than $270 billion each and every year. And that's just because of the use of Anywhere IT; it doesn't include the real estate savings such as those cited by IBM because of Anywhere Workers who don't require an office.

In short, if you don't become an Anywhere Enterprise, you'll be competing with businesses with IT cost structures that may be only 10 percent of yours. Good luck with that.

Anywhere Enterprises Are Greener

According to a report by the U.S. Environmental Protection Agency (EPA), from 2000 to 2006 the energy consumption of U.S. data centers doubled from 30 billion kilowatt hours per year to 60 billion. When it's running at full scale in 2011, a large facility such as Google's data center at the Dalles in Oregon will consume more than 100 megawatts, or about as much power as a small steel mill. Yet most of that power is used very inefficiently; according to a 2008 IBM study, one watt of business application computing requires an average of 27 watts of power and cooling at the data center. Unused servers and capacity make those numbers even worse. (See Figure 8.5.)

Anywhere Enterprises will turn these numbers around. Because servers don't consume much more power when they're working hard than when they're idle, combining many virtual servers onto one physical server saves power. But even better is the fact that cloud-based servers—those that are shared not only among many workloads, but among many companies—can be reallocated when they aren't needed. When Nordstrom finishes its Christmas selling season and doesn't need all the temporary server capacity until the next surge, those servers can be reallocated to help tax preparer H&R Block gear up for April tax time, all with the click of a mouse. In essence, cloud computing servers get recycled rather than wasted.

The green results of Anywhere IT are significant. According to the same EPA study, best practices such as those used by Anywhere IT can reduce U.S. data center power consumption by 60 billion kilowatt

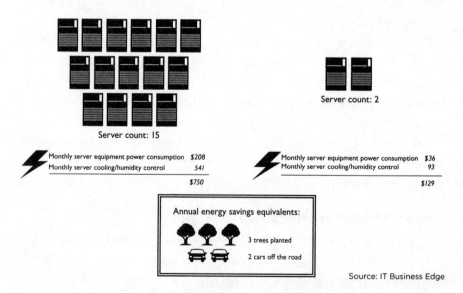

Server count: 15

Monthly server equipment power consumption	$208
Monthly server cooling/humidity control	541
	$750

Server count: 2

Monthly server equipment power consumption	$36
Monthly server cooling/humidity control	93
	$129

Annual energy savings equivalents:

3 trees planted

2 cars off the road

Source: IT Business Edge

Figure 8.5 *Virtualization and the greening of IT.*

hours per year by 2011, and in the process, eliminate the emission of 38 million metric tons of carbon dioxide. Observant readers will note that 60 billion kilowatt hours per year is roughly the same amount of power consumed by all the data centers in the United States in 2006. Combine these power and emission savings with factors such as reduced commuting pollution by Anywhere Workers who can do more work from home, and we can see that Anywhere Enterprises have more to offer than just cost savings.

If your enterprise cares about green business and its corporate social responsibilities, consider this: Much of the world is already adopting cap-and-trade policies that tax greenhouse gas emissions. Imagine competing against Anywhere Enterprises that have substantially lower taxes because they use their networks in smarter ways.

Anywhere Enterprises Are Dramatically Different

Amazon's Werner Vogels notes that before the firm's own cloud-based computing program was launched, "Only about 30 percent of our time, energy, and dollars, in IT was spent on differentiated value creation. That meant that 70 percent of our time, energy, and dollars, was spent on undifferentiated heavy lifting." Amazon's goal for itself was to reverse that split. Because the investments it made to do that for itself allowed it to offer the same services to other business, it made a dramatic shift for an enterprise most people would describe as a retailer.

This same transformation came up again and again with the executives we interviewed for this book. Anywhere Enterprises don't just improve or outsource their business applications; they reinvent them by using mobility and connectivity as their catalysts. The business results from this reinvention often are astounding. It's like having an eBay for work: When you have work to do, you find someone else who wants to sell you that work, all made possible by ubiquitous connectivity.

"Ubiquitous connectivity makes work a verb, not a noun," says Larry Weber, chairman and CEO of W2 Group. "Some things won't ever be 'done' anymore. We will keep continuously going, building, evolving, adding on, and changing them."

So a new Big Bang is coming—the Anywhere Revolution. The emergence of the Anywhere Network is changing not only all of us, but the enterprises that use it. Your business will either become an Anywhere Enterprise or end up competing with one. It's only a matter of time.

In that spirit, the next part of this book examines two central questions: First, how to determine the urgency with which you should be creating your own Big Bang; and second, what tactics will help your enterprise profit from the effort. We set out key steps you can take to position your business by establishing just how Anywhere you need to be to survive and flourish in an Anywhere world.

PART IV

Profiting from ANYWHERE

Chapter 9: How ANYWHERE Do You Need to Be?

"Mobilization of the world will continue exponentially. You need to be extremely knowledgeable about what effects that will have on your business."

—Rob Conway, CEO, GSM Association

Have you heard this one? Someone tells you she knows the secret to all comedy in the world, and invites you to ask her what it is.

"OK," you say, "so what's the secret to . . ."

"Timing," she interrupts.

And so it is with business. Sometimes success is about having umbrellas for sale in a sudden downpour—being at the right place at the right time. In the last few chapters, as we explored the implications of the continuing advance of connectivity in our lives, we've seen a lot of commercial profit opportunities ahead. Sounds easy enough, but when do you get those umbrellas set up on the street corner?

Move too fast and you might end up with the LG refrigerator, featuring a full Internet browser on the front door before most of the U.S. market had broadband access at home. From where we sit now, seeing the inevitability of connectivity's expansion, it was certainly the right idea conceptually, but it was years too soon for the market's acceptance and before the industry itself had figured out the appropriate role of connectivity in the device.

But wait too long to move on new market and product opportunities, and not only do you miss the opportunity for profit, but your business could be completely marginalized by competitors who do seize the day. Which U.S. automotive manufacturer doesn't wish they'd introduced the Toyota Prius, which as of this writing was continuing to generate waiting lists of hopeful buyers at a time when Detroit can hardly give its cars away?

The purpose of this chapter is to help you decide how quickly you need to move, and how quickly you *can* move, to profit from the Anywhere Revolution. We take a simple look at your markets, products, people, processes, and technology, with the goal of figuring out how to get these to come together at the right time to help you make money.

Put Your Business to the Test

Your goal is to get your Anywhere readiness as an enterprise to coincide with the Anywhere opportunities that are coming your way.

The first thing to consider is the definition of our key term "enter-prise." What is an enterprise? For our purposes here, let's agree that it's a group of people who have identified a set of *markets* and some *products* to provide for customers in those markets, using a set of *activities* and *technologies* to do so. An enterprise can be a nonprofit, of course— even a parent-teacher association has customers of a sort. We look at each part of that definition and examine its role in the need to adapt the business to the Anywhere Revolution.

Let's start by picking a point in time to aim for. Since many firms are comfortable with a five-year outlook, we'll use that. What we need is a sense of how Anywhere will affect our businesses by that time. To deter-mine how Anywhere you should be by 2015, there are two sets of issues to consider. One is about your markets and products, and the other concerns your organization and how it gets those products into the hands of your customers. If your customers and products aren't changing rapidly, you won't want to push more Anywhere offerings into the marketplace than your customers are willing or able to adopt. The external competitive pressures on your enterprise will be milder, and your near-term opportunities to profit from Anywhere will be more internally oriented. Conversely, if your customers are adopting Any-where activities quickly but your organization isn't, your efforts might be overtaken by events. You could become roadkill on the highway to Anywhere if you don't move faster on external priorities.

Looking at both these possibilities will help us estimate whether your ability to profit from Anywhere is *ahead, in sync,* or *behind* its evolution in your markets. So think of this chapter as a test, a sort of Anywhere quiz. I'll talk you through the basic questions you need to ask, explaining their purpose and the significance of your potential responses. The questions that appear in the heads that follow are also presented in the illustrations in a format that will enable you to keep score as you go. The questions are meant to be relatively easy to answer; if you don't know the answer, make your best guess. (If you can't even guess, then your first order of business in helping your firm profit from Anywhere may be to learn more about the company!) You'll give your firm some points for each possible answer, and we'll total them up at the end and see what it all means.

Customer and Products

Question	Answer choices	Points	Your score
1 **Markets:** How Anywhere will the key markets for your products and services be in five years?	**A:** 100% +	50	
	B: 33 to 100%	35	_____
	C: Less than 33%	20	
2 **Product Type:** How Anywhere can the products and services in your industry become?	**A:** Anywhere bits or services	50	
	B: Anywhere-enhanceable atoms	35	_____
	C: Anywhere-aided services	20	
	D: Anywhere-trackable atoms	10	

Total _____

Plot this on the y axis in Figure 9.3

Source: Yankee Group

Figure 9.1 *Anywhere Quiz: Questions 1 and 2.*

Question 1: How Anywhere will the key markets for your products and services be in five years?

Our first question calls for an assessment of how quickly your key markets are moving toward Anywhere: how pervasive the broadband network infrastructure will have become in five years in the parts of the world where you do business. (See Figure 9.1.) To figure this out in a simple way, we can use Yankee Group's Anywhere Index, which we examined in Chapter 3 when we asked, how fast is Anywhere happening? Using current penetration data and constantly tested assumptions about growth rates by region, Yankee Group's analysts forecast the likely growth in penetration ahead.

Consult Figure 3.4 to revisit our outlook for some of the regions of the world by 2015. As you make your assessment about the outlook for your firm's key markets, you should consider, on a proportional basis, where the people who are going to touch your product or service will be around the world. Think about both your customers and your customers' customers. Since most businesses get 80 percent of their revenue from 20 percent of their customers, you can focus on the regions of the world that matter most to you. If your top 20 percent have a pretty high Anywhere Index, you can assume that if you meet their Anywhere appetites, you'll be appropriately positioned to benefit as other markets follow suit.

Keep your assessment simple and focus on the markets that matter most to your organization. Give your firm 50 points if your key markets will be fully Anywhere by 2015: for instance, Japan, Italy, Sweden, the United States. People and organizations in these parts of the world will have the most advanced and reliable appetites for Anywhere products and services. If your core markets will be in the transforming range by then, meaning a mixed market where some areas have a pervasive broadband experience and others don't, score that with 35 points. Finally, if your core regions will be only emerging (fewer than one broadband line for every three people), score that with just 20 points.

Regardless of how your markets scored, if their Anywhere Index is dramatically different from where you are and where most of your organization is physically located today, you should also make sure you understand the very different issues by region. We've discussed how Anywhere will allow our businesses to shed the baggage of physical location and become more virtual, but for today, most firms are very clearly tied to particular locations, which creates an almost invisible bias in how we see the rest of the world. Dr. Nicholas Negroponte is the founder and chairman of the One Laptop per Child nonprofit, whose mission is to bring connected computing to emerging markets. In describing commercial PC manufacturers' efforts to serve those markets, he says, "They all assume that a child will be plugged in with an AC adapter. That's not real." Less than half of the general population in some emerging markets has regular access to electrical power; many instead rely on diesel generators or car batteries to power the devices they use. Do you truly understand how advanced or how behind your markets are? What could you do to give everyone in your firm a better sense of what connectivity is like where you sell?

Question 2: How Anywhere can the products and services in your industry become?

Everything that firms in your industry make or provide—whether they are widgets or people—is affected by the rise of global connectivity, but in very different ways depending on what they are. So in our assessment of how fast your firm can and should move, let's look next at how the kinds of products you offer have the potential to be changed through connectivity's ubiquity.

To fully understand the connectivity potential of your offerings, we need to consider the ability of ubiquitous connectivity to intrinsically transform the essence of your product or service—what it does and how it does it. And in that core sense, it's about your products and services as a class—considering their *potential* for connectivity, not whether your firm, or any other in your industry, has yet added connectivity to it.

Remember that you can't control the pace of the Anywhere Revolution or anticipate its direction. Even if your firm chooses not to apply connectivity to what you sell, you should still assume that your competitors are looking at how to incorporate connectivity into their strategies. Your only real option is to consider your products' and services' core connectivity potential as you plan when and how to make changes to your own operations.

The answer to this question divides the world of products or services into four categories, depending on how connectivity can affect them.

Anywhere Bit Products

Nicholas Negroponte's seminal 1990s book *Being Digital* split our world into "bits" and "atoms"; that's a useful place to start. "Atoms" are things we can touch, things that can't be squished into a cable and sent to an electronic device somewhere else. But if your product is, at its core, composed of "bits" rather than atoms—that is, if it is digital at its essence, or it's a digitally provided service, then the impact of Anywhere is total. Why? Bits can appear nearly instantly anywhere that the network itself appears. This presents huge urgency and huge risks. Nowhere is this easier to observe than in the music industry's battles with the onset of digital music in recent years. Music, like water, is very difficult to contain; you can't easily stop it from flowing across borders into other digital devices. This is one of the defining challenges of bit products in general.

Given the advances in digitizing previously analog content—photos, movies, X-rays, books—many more products than just music are now nothing more than bits at their core. Money is now an essentially electronic representation, and paper currency is less and less prevalent in our lives. Financial services firms, which have already moved to implement electronic funds transfer in the middle of financial transactions, are now moving to total electronic transactions.

Give yourself 50 points if your core products and services could potentially be rendered as bits that go Anywhere. If your product or service is bits, you'll have to move to Anywhere with speed. On the other hand, if your product or service is irreducible to bits and must be composed of atoms (something you can touch), you may be on a different timeline.

Anywhere-Enhanceable Atoms

Let's look at the second group; these are products composed in their essence of atoms, but atoms whose core functionality can be enhanced by connectivity. Remember our connected umbrella, made more "umbrella-ish" by virtue of its ability to tell us when we need it? Anywhere-enhanceable atoms are tangible things that could be more useful in what they do for their users with the addition of connectivity.

These can include the variety of consumer products we considered in Chapter 4, as well as a vast landscape of things in the business world that can take advantage of connectivity to send and get information, either with human users or with other things. For consumers, gaming devices such as the Xbox are dramatically enhanced by connectivity to other gamers and new gaming functions. In commercial applications, RFID technology provides the means for valuable assets to talk to the network about their status and to accept instructions.

One way to enhance a product's value through connectivity is with the collection of devices that get connected. In a truly Anywhere world, you might make a reservation for dinner on OpenTable which would automatically put the event in your calendar—which, living online, would appear on any device you select to access it. Your phone would remind you of the time, and you'd hop into your car with the navigation system automatically programmed for the restaurant. Once you're in the city, your onboard system will point out the nearest parking place and submit payment. When you get out of the car, you could be sent a text message with walking directions. The restaurant could be notified that you'd parked and would get your table ready. As you pass through the door, if you'd set your handheld's near-field communications capability to identify you to nearby devices reading it, the maitre d' would greet you by name.

Intelligence and value created by connecting things *to each other* for a threaded experience could ultimately add more value than connecting just one item. Score 35 points if your core product offerings are Anywhere-enhanceable atoms.

Anywhere-Aided Services

What if what your firm provides is a service of some kind? From call centers to corner dry-cleaning operations, the world's service sector is immense, totaling an estimated 62 percent of global GDP. A ubiquitous, capable communications fabric allows the performance of many of these activities to be moved to labor centers that are less expensive—witness the rise of outsourcing in India and other countries—or the consolidation of those activities into one location for simple efficiencies. Fast-food chains in parts of the world today already have moved their drive-through order-taking from one person in the restaurant to a call center located elsewhere. There, operators can field a high volume of orders and send the desired items to the right screens at the right restaurants with no interruptions to refill the in-store cup dispenser. Customer satisfaction with the experience tends to be higher, but the principal Anywhere profit payoff comes in the lower cost per order taken and the greater flexibility of in-store staffing. Axel Haentjens, the senior vice president of global strategy for network operator France Telecom, says, "All service businesses—banks, travel agencies, governments, more—need to worry quickly. Customers will not support not being served immediately. As they use these tools in their personal lives, they will expect the same in their interface with you."

Even when a service can't be performed somewhere else on the network, connectivity can transform the efficiency of the service. Think back to our Anywhere utility worker in Chapter 7, equipped with connectivity tools that increase the number of stops he can make every day and shorten the time at each stop due to the higher reliability of the information he works with during the visit. Give your firm 20 points if its core offerings are services that the Anywhere Network can relocate or assist.

Anywhere-Trackable Atoms

Our final product category contains those things composed of atoms, as with the second group, but that may not derive any potential added value in what they do for their users from the addition of connectivity. A bottle of olive oil, for instance, can't be any more oily with an RFID tag on it, connected toothpaste can't get our teeth cleaner, nor can literally millions of other things in our lives get better with Anywhere no matter how simple and inexpensive connectivity technologies become.

But this doesn't mean that connectivity, when added to them, won't revolutionize your business, or that there's no way to profit from the addition. If we care about them in some way as manufacturers or purveyors, we're interested in core information that makes them what they are—their freshness, their origin, their ingredients. And we certainly care where they are in the world. Thus this category is for products that can be tracked and made visible to the global network fabric using connectivity technology. Does this pallet contain a full load of Sheetrock? Did it leave the loading dock yet? Did it get wet en route? Once connectivity technologies such as RFID are sufficiently inexpensive and the software solutions to consolidate the explosion in data these will generate are mature, the processes to create and deliver most of the world's goods will have incorporated connectivity all along the way. Score 10 points if your firm's products are mostly atoms that could be tracked with expanded connectivity.

Does your firm offer a mixture of products and services across these categories? If so, do a quick bit of math to score them separately, and adjust the total, no more than 50 points, by the proportion each type represents in your revenue mix.

*   *   *

Now we're ready to move on to the final three questions. See Figure 9.2.

Internal to Your Enterprise

Question	Answer choices	Points	Your score
3 **People:** How ready for Anywhere are your people?	**A:** Our employees don't feel tied to specific applications to get their jobs done; new technology in the firm is absorbed with minimal disruption. Employees often try out new ideas on their own.	30	
	B: It's painful to roll out new initiatives in our enterprise, but we do it, either by top-down edict or by grassroots energy.	20	_____
	C: We don't try anything new until we know it's fail-safe; the rest of the world has to figure it out first.	10	
4 **Activities:** How Anywhere are the activities that help you make, deliver, and service your products?	**A:** We systematically review activities to reduce the costs of time and distance, and to improve access to information for all our employees and partners. We've already changed some activities and are committed to doing more.	30	
	B: We're learning: we look at delays in process or missing information, and we work to speed things up by introducing more connectivity.	20	_____
	C: We don't change how we do things very much.	10	
5 **Core IT platform:** How Anywhere is your core technology platform?	**A:** Our IT operation is moving to Anywhere: we use five or more types of network-centric solutions.	30	
	B: We're using or experimenting with two or three.	20	_____
	C: We're not working with any of these Anywhere technologies yet.	10	

Total _____

Plot this on the x axis in Figure 9.3

Source: Yankee Group

Figure 9.2 *Questions 3, 4, and 5.*

Question 3: How ready for Anywhere are your people?

Let's look at your own organization and the key factors that determine your readiness to take advantage of the Anywhere Revolution. Remember that we're aiming to get your firm ready now for a future point in time. This question and the two that follow will help assess whether your enterprise is on the right path and how aware and responsive it is to the future of global connectivity.

At its core, the secret to profiting from the Anywhere Revolution is a firm's ability to transform. Any enterprise can change; it happens every day through decisions to open and close factories, kill or launch products, buy or sell divisions. But real transformation—a pervasive adoption of new ways to work and new things to offer the market— can best be achieved with committed leaders. As you consider the Anywhere vision I've laid out, ask yourself whether your company's leadership understands the vision of Anywhere. Are the leaders curious about new technologies, and are they experimenting with them? If they're changing hand sets, trying out new things, that's a good sign.

Yankee Group analyst Zeus Kerravala puts it very simply: "Leaders should just use the new stuff. It shows their employees that they get it, and that helps their companies move forward." Cisco, the global networking technology company, uses telepresence extensively inside its own organization to make decisions with speed and efficiency. A Web-based dashboard tracks the total time and travel expense saved.

Consider the attitude of your own leadership toward such things: toward Twitter and social media, toward blogs and mobile applications, toward the idea of staff members working from home rather than coming to the office every day. In the past year, what experiments has your company's leadership led or encouraged that relate to technology development or connectivity? And how do they talk about change more broadly—are they focused on the future horizon or only the next few quarters?

But let's not leave all the responsibility with enterprise leaders. It takes more than a CEO with an iPhone to get an entire enterprise to understand how the expanding network will change the opportunity for profit. What's the rest of your organization's workforce like? Can the rank and file also adopt change?

At Research In Motion (RIM), the entrepreneurial Canadian firm behind the wildly successful BlackBerry, a single employee developed its mobile Facebook application. He was a Facebook enthusiast, working on his own time without any direction from above, and it took him about one day to develop the application. On a lark with few expectations, the company decided to make the application public in November 2007. In the first 16 months, the application was downloaded 5 million times. "Something like 60 percent of RIM's devices are now sold to consumers rather than businesses," says Yankee Group analyst Josh Holbrook. "That's a stark change from a few years ago. Handset device design is important, but you also need the applications that consumers want." The effort didn't garner RIM additional direct revenues, but it broadened the appeal of the product to resonate with Anywhere Consumers.

An enterprise with a future-facing workforce, top to bottom, is a powerful thing. But either leadership or a grassroots culture can help the enterprise move forward. The only scenario that won't work at all is when neither the company's leaders nor its workforce has a clear orientation toward change. Give your firm 30 points if it's responsive to change opportunities throughout the firm, 20 if it's less than perfect but has some ability to create change, and 10 if change is something the firm avoids at every turn.

Question 4: How Anywhere are the activities that help you make, deliver, and service your products?

Chapters 7 and 8 of this book show how the Anywhere Revolution can collapse time and distance to bring us all closer together in business, delivering information at the immediate point of opportunity. How much you need to be thinking about how this will impact your company probably depends on the extent of physical distance in the activities undertaken by your enterprise. If the work in your firm is very local—you buy food at a local farmer's market, and bring it back to a restaurant kitchen to prepare it, for instance—then the opportunity to incorporate more network-based thinking into your internal process is relatively limited.

On the other hand, if distance or time are factors in your activities—if you wait for parts from China, or if you deal with month-long or even year-long steps in the development of a design for a car, for instance—then network-based thinking to reduce latency and distance is not only a core profit opportunity but an imperative for you. "Latency is evil," says Paul Sagan, president and CEO of Akamai, a firm that accelerates Web content delivery across the Internet. "We've estimated the billions of hours saved by speeding up the Web experiences our customers offer. It's milliseconds here and there, but it adds up to over 4 billion hours saved and counting." You need to be thinking about gaps in time and space as being black holes of lost profit.

Finally, if your firm is very large or your supply chain/network is complex and if lots of other firms are involved, these are additional dimensions ripe for connectivity-based thinking.

Bottom line: The more entities, steps, items, kilometers, people, dollars, and days in your enterprise, the more opportunity there is for Anywhere technologies to help you reduce expenses and increase profit. Just remember, that same opportunity is probably available to your competitors. Enormous potential advantage will accrue to the firms that move first to collapse these gaps in time and space. Give your firm 30 points if it systematically reviews processes and activities across its endeavors to quantify the impact of time and distance; 20 points if you're just learning, considering how connectivity can save time through a pilot or two; and 10 points if core processes in the firm are cast in concrete, frozen by rules, regulations, or even just tradition.

Question 5: How Anywhere is your core technology platform?

Anywhere is about the integration of connectivity into everything we do. Nowhere is this more relevant than in the core information technology that a company depends on to do its work.

Since the introduction of the first computers into the commercial world in the 1960s, firms of all sizes depend every day on a core computing technology platform. Newer network-centric technologies will render today's IT operations and conventions almost unrecognizable to the companies of the future, even as soon as 10 years from now. It will most probably mean owning less IT equipment at the firm and depending instead on the external provision of IT capability by outside vendors. Remember those nineteenth- and early twentieth-century companies that had to provide their own power? Think about what a savings in complexity and expense you can achieve by reducing your dependence on computing hardware and software in house.

Information technology solutions that incorporate a persistent and capacious network are blooming rapidly; we itemize a number of them in Chapter 8. If your firm is already beginning to use these, the technology platform in your enterprise is becoming Anywhere. That's a big advantage because it suggests that your IT group knows the benefits, is building its skills with these technologies, and has had some success with getting leadership support for introducing them.

Is your enterprise investing in the change available to its information technology infrastructure? Score your firm 30 points if it's using at least five of the network-centric technologies we mentioned in Figure 8.2; score 20 points if it's begun using three or four, and just 10 points if it's anything less than that. The less your IT platform incorporates these elements now, the harder it will be to catch up because the internal team needs to master what they are and how to work with them. It's a big change in thinking, and it will mean moving away from past technology platforms—computers, applications, partners—and toward new ways of thinking about them.

Where Do You Stand?

Reckoning time: The purpose of these five questions has been to estimate how Anywhere your business is today compared to what the external market and the products you offer will require in five years' time. You should have a total for each of the two sets of questions of some number between 0 and 100. Plot each total on the two-dimensional box in Figure 9.3. There are four general ranges of outcomes.

Upper Right

If the two subtotals of the answers to the five questions we asked put your firm in the upper right of our Anywhere quadrant, it suggests that you're *in sync* with the evolution of your core markets and the capability of your products to benefit. Your firm's focus should be to beat out competitors by deploying Anywhere technologies before they begin using them. Think of Avis, pushing the rest of the car-rental industry to respond. Could you remake your industry's approach to costs by accelerating your adoption of connectivity? Focus on adding connectivity-based benefits to your products before your competition has the opportunity.

Lower Right

If your totals put you in the lower right quadrant, you could be *ahead* of the game. Your firm may be moving more toward becoming an Any-

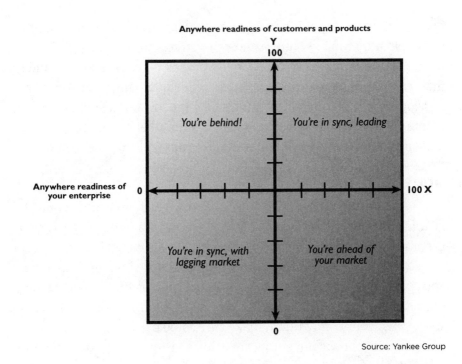

Figure 9.3 *Scoring your Anywhere urgency.*

where Enterprise than your markets may be ready for in the near future. In 2009, some Web sites serving lots of video across the network reduced their stream delivery to some parts of the world because they weren't yet able to monetize it by selling ads. Perhaps you're with a brand-new startup with no pre-Anywhere IT baggage, so you've been able to begin with Anywhere thinking from the start. A focus on the cost-reduction benefits of Anywhere will help leverage scarce capital while the external markets begin to match your pace.

Lower Left

If your score falls in the lower left quadrant, you're probably *in sync*, but with slow-moving customers and products. Neither your firm nor your external market is very ready for Anywhere. Regional manufacturing businesses operating in Latin America, for instance, have a limited broadband environment and don't need to make a fast move to Anywhere. They can save money by implementing asset tracking within their plant facilities using internal Wi-Fi networks, but the scarcity and cost of broadband in many Latin American markets make it less critical for them to expand the use of connectivity in their external activities and in their core IT infrastructure. If you are in this quadrant, your Anywhere mission should be to move your internal score toward the right quadrant before the Anywhere Index in your target markets rises. You need to anticipate and perhaps even drive your customers' moves to Anywhere.

Upper Left

If your score is in the upper left quadrant, your enterprise could be *behind* the Anywhere Revolution in a serious way. In this case, your first mission in profiting from Anywhere should be to help your firm develop a sense of urgency, getting everyone focused on this imperative. You might need to spend more time in the regions of the world where your customers are most Anywhere. If you regularly benchmark yourself against your competition, make sure some of that effort is aimed at understanding how competitors are using a fast-growing global network to reach clients and deliver expanded value. Focus your merger and acquisitions efforts on companies based in those regions or that bring you more Anywhere solutions for what you do. And stop investing in "somewhere IT," that is, solutions that tether workers to fixed places and constrain the ways they access your firm's critical assets.

* * *

We've used some simplifying assumptions to help you develop a working perspective of the urgency of your firm's Anywhere transformation. Next, we'll take a tour of your firm to figure out where your best Anywhere profit opportunities lie. But keep in mind that the secret ingredient, as with comedy, is timing. You need to plot the connectivity trajectory your customers and your products are on, and assess your firm's commitment to it, to determine whether you have the right level of focus. Whether you're ahead, behind, or right in sync, whether you're an individual contributor or a leader, whether your firm is large or small, there are lots of things you can do to profit from what's coming in connectivity.

Chapter 10 | How to Go ANYWHERE

"It would be a mistake to think there will be companies that use ubiquitous connectivity and ones that don't."

—Sriram Viswanathan, Intel

Where were you in the late 1990s? In 1995, the Internet browser start-up phenom Netscape had the most highly anticipated initial public offering ever to that point, with its first-day share price exploding from $28 to $75. In 1996, America Online, then the leading proprietary online service connecting consumers in the United States and elsewhere via dial-up phone networks, was exploding. I was an Internet analyst at the time, fielding calls from panicked company leaders. "Our board of directors is all over us about the Internet. What should we do?" they'd ask.

For many companies, what followed those early catalysts of the Internet's commercial maturation was a five-year period of largely chaotic and disorganized activities to try to seize the day: launching dozens of different Web sites run by disparate divisions or even lone employees, rushing into decisions to move too early to online commerce or supply chain initiatives with half-baked technologies and partners—and arguing, lots of arguing about whether the Internet was some kind of fad or an important change that would affect every firm in business in some fashion or other.

Ultimately, of course, the verdict emerged: the Internet was going to matter to all of us. But during that time, millions of expense dollars were wasted and just as many in potential revenue were overlooked because of missteps, debate, and inaction.

Let's not do *that* again. In light of all we've discussed about the opportunities to profit from the further expansion of connectivity, the question now is simply this: What's the best path forward? In this chapter, we first recap the expanse of Anywhere profit opportunities, then focus on your business to find the ones that are relevant to you, and, finally, set some priorities for determining which to tackle first.

Where's the Beef?

As I write this, businesses around the world are dealing with another massive change: a challenging global recession that has bankrupted some of the biggest business names out there. Many companies have been cutting back, and major industry players are closing ranks in an effort to keep their heads above water. New profit has been tough to find.

But Anywhere profit is there for any firm. You just have to know how to look at it. Remember those diagrams of colored dots, the ones that appear to have no pattern when you look at them in a normal way? If you change your perspective by putting on a pair of polarized glasses, you see a pattern, a number or a letter. It's the same with Anywhere. If you change your perspective, if you start looking at how connectivity will change your products, your customers, and your organization, you'll begin to see things differently. (See Figure 10.1.)

For any business, increased profit of any kind—growth in the bottom line—comes in just two ways. You can increase your revenue while keeping your expenses at a constant level, or you can lower your expenses while keeping your revenue stable. Or, if you're really ready to embrace the Anywhere Revolution, you can do both.

Let's start by recapping the top-line opportunities. For most businesses, there are two ways to increase revenue. The first option is to secure more revenue from each of your company's existing customers. You can increase their "lifetime value" to your organization, or sell them more things; either expands your role in their lives.

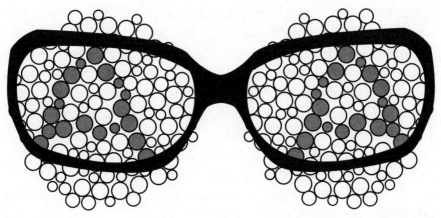

Source: Yankee Group

Figure 10.1 *Anywhere vision.*

You can maximize the value of customers you already have by adding connectivity to existing tangible products. In Chapter 4 when we look at the emerging portability of experience, we provide a number of examples of products that could deliver expanded value thanks to the addition of connectivity.

These types of opportunities could come in the form of a service contract, committing to periodic updates, or enhancements over a period of a year or more for an extended fee at point of sale, or else in a higher up-front price, bundling in the cost benefit of updates to the actual price of the product. Given the rising tide of consumer and business interests in sustainability, Yankee Group sees opportunities for firms to incorporate intelligent power on/off features for high power-consumption devices, such as air conditioners, washing machines, and cars.

To take advantage of our Anywhere future, businesses should expand services and information delivery over the global network just as Yamaha does with its connected piano. Its connection to the network means that enthusiastic pianists can access a growing database of music at the actual point of use.

Many service providers will need to deliver directly to users anytime and anywhere. In the travel sector, many airlines see the need for consumers to get information and services from mobile devices. Among other airlines, Singapore Airlines, Delta, and KLM have all recognized this and have facilitated mobile device access to their Web sites for check-in and flight status tracking. And brand-new Anywhere consumer services are emerging to meet the needs of people moving around. Buzzd is a service that lets mobile users in the cities it serves know what's happening in town at any given moment, bringing location-based guides to the BlackBerry and other smartphones. "Mobile devices are the only platform where space and time truly converge," says Nihal Mehta, founder of location-based mobile city guide Buzzd and a long-time mobile entrepreneur. "Consumer location-based services are just starting to take advantage of this, where the phone can detect this relevance and apply it as utility."

With our Anywhere glasses on, we can see people untethering from specific locations and habits, newly able to use information and support delivered instantaneously on whatever device is convenient—a mobile phone, an electronic reader, a display in their car, a kiosk they pass by, whatever it happens to be.

Your second option for increasing revenue with Anywhere involves gaining more customers. Your success at adding new customers in existing markets will depend in part on the speed with which your firm adds connectivity to what you do for customers compared to your competition. You should already be aware of what your competitors are doing, but you'll need to look again wearing Anywhere glasses.

Beyond winning new customers within existing markets, you'll be able to take advantage of the billions of potential customers in new markets that are now joining the global digital network fabric. According to the ITU, the telecommunications standards organization within the United Nations, fewer than 2 percent of sub-Saharan Africans had mobile phone access at the turn of the century. Less than 10 years later, mobile phone penetration has shot up to 25 percent in the same area. With new consumers continuing to sign up for their very first mobile phones at stunning growth rates, we're going to see the results of this continued expansion in less than three years.

When it comes to targeting these new markets, keep this in mind: most of them have dramatically different socioeconomic conditions from markets with more mature communication services. They will bring customers to the Anywhere Network with lower incomes, less education, and different literacy levels and lifestyles. Their network experience may not look like yours. And other parts of the common infrastructure of our more developed world may still be largely missing in these newer markets. You'll have to monitor the establishment of these basics, including power, water, and sanitation, and even the availability of street addresses. And these new consumers of the Anywhere experience will be coming to your network almost completely limited to the experience you can provide on a mobile hand set, not a PC—so screen size, means of input, amount of local processing power, and all other interaction elements will be substantially different from what they are for many of the customers you already serve.

As daunting a set of challenges as these markets may present, they will offer attractive customers to some firms. Western Union recognized its unique opportunity to be a leader for the millions of people around the world without access to basic banking services, such as immigrants in the United States and vast portions of the populations of regions such as Africa. It now offers a safe way for people to send and receive money even without a bank account, with payment transmission to a mobile phone that the hand set's owner can take to any licensed agent for conversion to cash.

Vanu Bose, a network technology entrepreneur, describes India as an exciting marketplace teeming with new Anywhere Consumers. While 70 percent of India's population is rural and 80 percent of that group is illiterate, he says, "They have the same basic demands and desires for information and entertainment as anyone else does." In the earlier days of mobile network rollouts, they weren't considered attractive as potential customers, and thus frequently there was no simple message service (SMS) on offer in less urban areas. But with the advent of mobile data services in the region, these same consumers now download Bollywood film clips and highlights from cricket matches at phenomenal rates, more often in some cases than their urban counterparts who have access to those media through additional means.

IFFCO, the Indian Farmers Fertiliser Cooperative Limited, has served the rural agrarian economy in that country with fertilizer and other products for over 40 years. It joined with mobile network operator Bharti Airtel to provide access to vital agricultural information via simple voice messages. The two firms share office space, and IFFCO serves as a distributor for Airtel products and services, giving it a new source of revenue, but with the added value of IFFCO's daily messages about weather, tips on caring for livestock, recommendations of insecticides, and expected availability of water in local irrigation canals. The operation is now active in more than 15 circles (regions in India) and has acquired over 1 million new customers for Airtel since the program's inception. The commercial impact for IFFCO is the additional income through commissions on sales of mobile devices and related services, along with rents it charges Airtel to allow the operator to put cell towers on its properties.

At the top line of your business, Anywhere profits can come in across the board—new products, new services to existing customers, and new customers in current and future regions. What are the best new products and services you can offer? What are the new customer markets within your most immediate reach? Prioritize your Anywhere opportunities to develop a path forward at the pace that is right both for your business and your customers.

Profiting from Anywhere Cost Savings

To see the overview of Anywhere profit opportunities on the expense side of your business, channel the mind of your CFO and consider potential savings in two broad categories: in the costs associated with creating your products and services (bettering your gross margins) and in reducing the expenses of your core business platform (lowering both capital expense, which consumes cash and weighs on the annual budget in the form of asset depreciation, and ongoing overhead in the form of operating expenses like building rent and software maintenance).

Starting with the cost reductions available on the gross margin side, where the business's core people and activities to produce its products take place, there are three dimensions to Anywhere profit: *reducing time*, *reducing complexity*, and *increasing agility*.

Reducing Time

Most business processes can be simplified by "lighting up" assets in the operation to let them report on their location or condition, thus saving steps. John Halamka is the CIO of CareGroup, a cluster of seven hospitals in Massachusetts. "It's been estimated that nurses spend about 20 percent of their day looking for things," he says. "We have 5,000 pieces of medical equipment scattered over 2 million square feet of hospital real estate. Someone will ask, 'Where is the nearest ventilator?' We put active RFID in all our facilities over the past year. Now we can say, 'Here's the thing you're looking for.' It's a big time-saver."

"We call it *unified asset visibility*," says Gabi Daniely of AeroScout, which supplied some of the technology in the CareGroup solution. "It's not only about tracking. I don't just want to know where it is. I want to 'see' the asset. I want to know, for instance, that I have less than five pumps in the storage closet. I want to know that I'm missing a certain part."

Reducing Complexity

Anywhere technologies can eliminate steps from complex operations. Michael Saucier, CEO of Transpara, a company that provides operations intelligence software for the utility and process manufacturing industries, tells this story: "We had a client with a fire at an offshore drilling platform. A remote engineer got an automatic notification on his Black-Berry, clicked on a live link, and saw some key stats about the status of the fire and the circumstances. The reduction in steps to notify someone of the problem saved them millions of dollars in getting the fire out faster. In the old process, there would have been phone calls to alert someone who would then go into the office to look into the problem. Edge decision making unlocks hidden potential in organizations."

The potential of Anywhere technology to reduce complexity extends to the means by which your firm acquires customers. So let's not overlook the long and sometimes fuzzy process you go through to create awareness, consideration, preference, and purchase of your products and services by your customers. Sir Tim Berners-Lee, widely credited with the invention of the World Wide Web experience we enjoy today, talks about the move the world is making to "pixel wallpaper"—the arrival in many new locations of digital screens that replace paper billboards and signage. Whether they're on taxicabs, in elevators, at bus and train stops, or at checkout counters, digital screens connect to the Anywhere Network to get the bits they display.

Rich Reiter with Bloomberg News explains the firm's approach to expanding its reach to mobile phones. "The company's core service is for its professional subscribers. And there is a mobile component for them. Now there is also a free consumer mobile platform that serves as an entry point to learn about Bloomberg and to get news. Mobile users get exposed to our data, then perhaps subscribe to the professional service. We're developing advertising and sponsorship on the platform to offset the cost to do it." As screens proliferate, the firm has new decisions to make. "We want Bloomberg News to be everywhere, but we have to be thoughtful. Could you wake up to a Bloomberg stock quote on your Chumby? Yes, but it's not necessarily for our core audience. But the Kindle, that's our audience." Which devices and screens will reach yours?

Increasing Agility

Finally, Anywhere allows businesses to be more agile, to respond faster to changes in demand. Confounding our goals as businesspeople to plan, many aspects of our businesses have a fundamental "bursty" nature—they experience rapid rises or drops in demand. Even assuming you could correctly anticipate rises in demand, preparing to meet that demand can saddle your firm with high fixed costs—labor, real estate, equipment—that aren't easy to shed if or when you don't need them any longer. Modern businesses operating in a global arena are more vulnerable to these phenomena than ever before. Maynard Webb at LiveOps asks, "What company doesn't want instant access to resources to help it ramp up quickly to meet demand, instead of losing quarters of growth? What company doesn't want the ability to dial those resources back down again if they run into trouble?"

Anywhere services like cloud computing allow firms to avoid investing in advance of demand but still be able to respond efficiently to it. Animoto is a Web service firm that lets users upload personal photos and images that it turns into music videos for them. The service was enjoying a reasonable level of success with about 25,000 customers—until the launch of an Animoto Facebook application in 2009, when it went to winning 25,000 new customers *an hour*. In just one day, the number of servers it needed to support the explosion in customer demand went from 50 to 3,500. Astoundingly, it managed to scale the service to meet this demand in three days without buying a single piece of hardware or having to create its own computing, network, and storage infrastructure. By renting infrastructure, it was able to meet its needs without spending on additional hardware. "No venture capitalist would have given them the amount of cash they'd have needed to provision themselves directly with 3,500 servers," says Amazon's Vogels. "So they would have been woefully underresourced for the surge in demand and would have missed the opportunity if they'd tried to do it on their own."

Moving from gross margin improvements to fundamental operating expenses and capital outlays, businesses of all types and sizes have the opportunity to enhance their efficiency with Anywhere technologies. You can reduce both hardware and software capital expenses through expansion of the pervasive network; savings might come from cutting down on the number of servers by virtualizing them, by reducing power consumption, by reducing the number of new PC purchases, by allowing employees to use their own PCs, and by moving to software-as-a-service applications like Salesforce.com. "That's only part of the story," adds Salesforce.com CEO Marc Benioff. "Developing new applications on a platform like ours is much faster. Cloud computing platforms will transform the CIO's job, freeing up resources and democratizing skills. It will deliver better security, transparency, and control than client-server computing without all the burdens—and at a fraction of the cost."

More and more businesses are recognizing savings through the application of advanced network-centric technologies. BAA, the world's leading airport company, employs more than 12,000 people, operates seven U.K. airports, and manages contracts at several major international airports. Its IT infrastructure included over 10,000 employee computers and over a thousand servers, amounting to a major support burden and hindering the IT team from creating new applications to drive business value. By moving to a consolidated network-based solution for providing access to its critical business applications, over 8,000 of the firm's employees can now reach over 200 of the company's business applications through any Web browser on any computer in the world. "The cost per seat for installing, managing, and maintaining users has been cut by 28 percent. Logged calls to the service desk are down by 11 percent, and second-line desktop support calls have fallen by 43 percent," reports Tim Matthew, technology services manager for BAA.

Over the past 20 years, many large firms with a mantra of constant expansion have assumed a "flag-planting" approach to the globe that has led to opening multiple locations to manage their global workload. Their leaders have felt the need to be everywhere, which meant securing and supporting an ever-expanding list of physical locations. With Anywhere technology, however, businesses can reduce expenses from multiple enterprise locations while still staying close to their customers. These days, you can be closer to your customers online by using the Anywhere Network.

You can also reduce the expenses you incur by operating your enterprise's own proprietary network. "If there's one uber-concept in all this," says Mark Templeton, CEO of Citrix, "it's that the Web will be the thing that enables the final set of silos to tip over in the distributed computing world. And the impact is that organizations need a plan to get rid of their privately-operated networks. They serve no strategic purpose anymore. The costs of securing and managing them . . . it's all wasted. A fully public open network is what they should move to."

Your Anywhere Profit Audit

Having taken a CFO-style look at where Anywhere profit lurks in a company's budget, it's time to consider how you're going to get at it. No two firms are alike, of course. I suggest that you start by doing an audit, taking a virtual tour of your business to determine exactly where these opportunities lie. Figure 10.2 can be your guide; it's a list of questions your firm can ask itself in an imaginary (or real) walk through

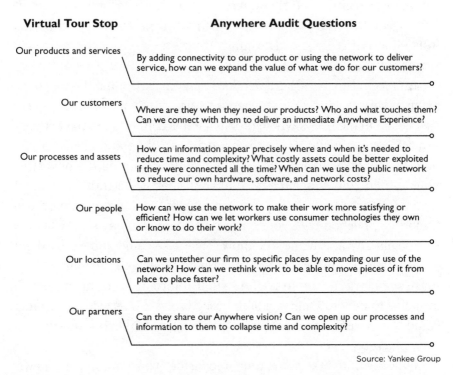

Virtual Tour Stop	Anywhere Audit Questions
Our products and services	By adding connectivity to our product or using the network to deliver service, how can we expand the value of what we do for our customers?
Our customers	Where are they when they need our products? Who and what touches them? Can we connect with them to deliver an immediate Anywhere Experience?
Our processes and assets	How can information appear precisely where and when it's needed to reduce time and complexity? What costly assets could be better exploited if they were connected all the time? When can we use the public network to reduce our own hardware, software, and network costs?
Our people	How can we use the network to make their work more satisfying or efficient? How can we let workers use consumer technologies they own or know to do their work?
Our locations	Can we untether our firm to specific places by expanding our use of the network? How can we rethink work to be able to move pieces of it from place to place faster?
Our partners	Can they share our Anywhere vision? Can we open up our processes and information to them to collapse time and complexity?

Source: Yankee Group

Figure 10.2 *The Anywhere profit audit.*

your operations. Many of them we've talked about already; those we haven't, we'll highlight here.

Your Products and Services

Consider the potential for an upgrade for your current products. Examine what could be achieved with the addition of a permanent or periodic link back to your company or to other products. One area to spend some time on is how you can expand the functional value of the product.

A grocery shopper who's hunting for pasta, tomato sauce, and parmesan cheese at rush hour probably needs the ingredients for dinner that night. Approaching the marketing of any of those items with this in mind—for instance, putting parmesan cheese in a rack in front of the tomato sauce—is an example of the *whole problem* philosophy. Online grocers such as FreshDirect are getting the idea and starting to offer suggestions to consumers buying products commonly grouped together. FreshDirect is even going a few steps further by offering a database of recipes and the ability for customers to buy every single item needed to make a particular dish with the click of one button, thus eliminating the need to jump from one page to the next or to search for the items independently.

Connectivity can be a tremendous asset in expanding the functional value of your product. What extended role could your product or service take on in the user's whole problem if it is connected to the network? What if people could use it to report in real time and check in with other assets or people to get or give real-time information or instructions? Could you reduce the need to go to specific locations to perform manual activities by adding connectivity? If you were granted a continuous connection with your customers day in and day out, with what you do already and what you know, what else could you do for them?

Another fruitful area to probe is how your own customers could make what *you* do more useful. What information from the user, if available to the product via a Web-based browser interface for instance, could the system retain and use—payment or authorization information, blood type, location, preferences?

On your tour, take some time to consider your product all the way through to purchase. Where are your customers today? You need to be able to reach your consumers anywhere—or someone else will. What is the decision process they go through? What are the steps along the way for someone to buy? Take a mental walk through the activities of your customers prior to their acquisition of your product. To get a sense of how Anywhere may shift customer behaviors, consider carefully how your customers experience your firm and its offerings today. (Connectivity will make this job simpler once you take the leap; Finnish do-it-yourself retailer Rautakesko uses RFID-tagged shopping carts to understand how its customers travel through its stores.)

Which other brands or organizations are touching your customer during that process? Think about payment methods, advertisers, transportation firms, media. What information could you exchange with them on the fly to expose your firm to these buyers? What locations do they go to? What digital displays or devices might they come into contact with there? Remember David Rose's favorite question: What would a terrific butler do just at that moment? What could you deliver to them at the exact time they are asking themselves a question?

Your Organization's Activities

Now let's move inside your organization. Imagine the steps you would take to show someone around your business. Think about activities that are repeated many times a day, week, month, or year. Look at all your core activities and ask these two questions:

1. Where's the latency in our activities?

2. What are these delays costing us?

To be fair, these questions aren't new. Many organizations study their processes carefully, looking to shorten delays and reduce errors. But our Anywhere glasses help us take this a step further by asking: When a network is all around us and devices can use it, which steps could benefit by having more immediate information? Where is that information today, and how could we deliver it precisely where and when it's needed? What other resources could be brought to bear on a problem or step in real time from the network to *expand* the data that aid us in decision making?

Your People and Places

As the global workforce changes with the annual retirement of older employees and the arrival of fresh faces, its attitudes about mobility and the technologies that support it, along with the role in their lives of the Internet itself, are shifting quickly. Appetites to relocate for the company are diminishing, while preferences for flexible work schedules and telecommuting are on a steady rise. Where could the global network bring them together more quickly to collaborate? How could the network allow them to shift work among themselves more easily? How could the network connect you with new candidates to take on work in your enterprise—people whose location or availability would have been too limiting otherwise? Will it save the company money if you allow employees to use their personal devices and preferred systems at work? What have your employees been asking for from the world of consumer technology that they're already familiar with or already own at home?

Locations will matter less, not more, in our operations in the future. Go to your company Web site to refresh your memory on all the physical locations around the world where you may operate. Consider the ongoing commitment of expense in real estate, the hard assets at each of those locations, and the efforts involved in maintaining those venues. How can you review your commitment to physical presence in that context? Are there somewhere IT technologies, such as desktop computers, that wed you and your employees to a particular location? What investment could you make to untether your employees?

Your Partners

All firms have critical relationships outside their walls that are vital to what they do: suppliers of parts, distributors of their product to the market, agencies that develop advertising—these and more comprise your firm's key partners. Consider the impact of these partners on your Anywhere plans. Are your partners as committed to Anywhere as you are? As you form new Anywhere plans for your products and activities, can they do what you want them to do, or will you have to select new partners with a connectivity-centric outlook?

Since the knife cuts both ways, you should also consider how well you work with your partners and how easy you are to do business with no matter where they are. Are you putting the Anywhere Network to work to make it easy for them to give you what you want and for you to give them what they want? The core challenge in extracting Anywhere profit in partner activities is making information exchange more immediate and more transparent; that is, opening up process and data to them over a public network accessible by any device at any time.

Your firm's technology partners—the companies that provide you with computers, software, networks, and technology services of all kinds—offer a rich area of Anywhere opportunity. The total expanse of those opportunities would take another book to do them justice. But simply put, if your firm has relationships with communications sector firms and you're close enough to that part of the business, then you have some great candidates for your profit audit. Here are some top-level questions you might want to ask them; for more, see the book Web site at anywhere.yankeegroup.com.

Here are questions to ask network providers:

1. Can you offer us a seamless experience between both wired and wireless networks?

2. What's your network's IQ? How are you expanding access to location and other network information through investments in network software systems?

3. How easily can we "burst up" and drop back with our use of network services based on fast-changing demands on our firm?

4. Are you building out your networks to provide sufficient capacity for our needs?

5. Will your network reach into the regions of the world we expect to expand into as their Anywhere Index rises? What partnerships extend your reach? Do those networks meet the same standards we have for yours?

6. How can we secure our activities on your network? How can you help us assess security risk?

Following are some questions to ask communications hardware, software, and related services providers:

1. How can you help us unify our communications to provide pervasive reachability to our employees and partners?

2. What are your plans for migrating to higher-bandwidth network technologies?

3. As we expand our use of multimedia in our communications with customers and partners, how will your technology support us?

4. How can your services be device-agnostic? That is, will using your service avoid locking us into a certain manufacturer or device type?

5. How Anywhere are the experiences you offer us as customers: your mobile Web site, your customer care solutions?

Prioritizing Your Anywhere Profit Opportunities

By wearing your Anywhere glasses on the virtual tour of your firm, you should have been able to identify quite a few opportunities to profit from the emergence of ubiquitous connectivity. But you can't do it all at once; nor should you try. With multiple opportunities, you have to choose the best and most efficient. How do you decide which of these opportunities to pursue and in what order?

There are lots of ways to evaluate technology projects. My advice is to keep it simple. Start by making a list of the ten or so connectivity opportunities you identified in your Anywhere audit that merit more rigorous research and consideration. Then apply the six criteria I list in Figure 10.3 to each one. Think fast; there's plenty of time later for the more detailed assessments that most enterprises ultimately require to green-light new technology. For now, just use one, two, or three checkmarks for each dimension based on how much positive potential the project affords, so that by the time you finish, each opportunity has a total of up to 18 checkmarks reflecting how suitable it is for your enterprise to tackle now.

Anywhere Profit Opportunity	Measurable ROI?	Competitive Advantage?	Internally Feasible?	Externally Feasible?	Makes Us Better at What We Do?	TOTAL

> Substantial value for our enterprise in this dimension
>
> Reasonable value
>
> Little value

Source: Yankee Group

Figure 10.3 *Prioritizing Anywhere profit opportunities.*

How to Go Anywhere

Among the many Anywhere thinkers we interviewed for this book were some who are successfully implementing connectivity solutions in their own firms and with their customers. Here are their words of advice on how to profit.

▶ How Should You Set Priorities in Searching for Anywhere Profit?

"Don't shy away from the basic trend toward connectivity. Focus on rich availability everywhere. Jump into the technology and provide things that are reasonably close to state of the art."

> —Axel Haentjens, senior vice president, global strategy,
> France Telecom/Orange Business

"The most important thing to do is to experiment. The good news is that Web-based applications, particularly those that let you reach customers, are easy to implement, and you can experiment without a lot of risk."

> —Ron Sege, president and COO, 3Com

"Seek to shed private corporate networks and devices. Calculate the true cost of plugging someone into a network that you own versus the public Net, and work to change to the latter. But don't see it as all-or-nothing; if you can get 20 percent of your workforce off a corporate network, look at the impact you can get from that."

> —Mark Templeton, CEO, Citrix

"Be careful to build connectivity experiences that are high in value, that are not just cool technology."

> —Dr. John Halamka, CIO, CareGroup

"Look at your consumption, your activities. How will you use mobile technologies to expedite, to optimize, to make what you do more efficient? Think about the question on a carbon footprint basis."

> —Rob Conway, CEO, GSM Association

▶ Where Should You Start in Your Anywhere Strategy?

"Solve your connectivity opportunities on the customer side first—you'll need that to stay competitive. Evolve internal processes to maintain productivity."

> —Terry Stepien, president, Sybase iAnywhere

"Start with this question: Is there a cloud-based model for doing this work that is simpler, that could give us 80 percent of what we need? Move to something simpler, faster, cheaper—and look closely at what exceptions you really need."

—Maynard Webb, chairman and CEO, LiveOps

"Don't start purely on cost savings alone; that shouldn't be your primary goal. And don't start with the most complex thing; you could get a real mess. Finally, start with areas where your partner ecosystem has some expertise and can work with you."

—Doug Hauger, general manager, business strategy,
Windows Azure, Microsoft

"Be curious and think creatively. Connectivity costs have come down, and the technology is mature. The potential situations where you can apply asset tracking for instance are immense. Think about the savings from knowing where everything and everyone is all the time."

—Are Bergquist, CEO, Matiq

▶ How Should You Lead Your Business through This Transformation?

"Retire and make someone else figure it out! Seriously, though: split your time between execution over the next five years and planning for the long term. And get some kids in the company to help worry about that longer term. Meanwhile, work out where you can be agile. The things that will be in our hands for the next three to five years are relatively well-defined. What services can you put on them?"

—Mike Muller, CTO, ARM

"Partner, partner, partner. Because of connectivity itself, where everyone is talking to each other, so many opportunities to develop companies and ideas, one company can't do it all. Partner with many firms, large or small, to accomplish your big picture."

—Steve Haber, president, Digital Reading Division, Sony Electronics

"You need to change how you think about human capital and resources. The workday—punching the clock, being on-site—will become less relevant. If you don't offer flexibility, you won't get the best talent."

—Dan Hesse, CEO, Sprint/Nextel

For each Anywhere profit opportunity you identify, ask yourself how well it can do the following.

Provide Measurable ROI

The Anywhere business opportunity is, after all, about increased profit. There's no need to pitch the launch of Anywhere projects inside your firm on the basis of warm fuzzy feelings; adding connectivity to products and processes has to be one of the most measurable endeavors a company could undertake, just by its very nature. You should be able to estimate the financial impact of new customers, product up-sells, or time or equipment saved for any project you consider. If it could be game-changing in terms of your firm's gross margins, give it three checks. Otherwise give it two or just one check, depending on your sense of the proportion of its profit contribution relative to other non-Anywhere projects.

Gain Immediate Competitive Advantage for Your Firm

How could this project or the insight it produces help you gain competitive advantage? Can you gain market share by the addition of connectivity? Lower your pricing to customers from a manufacturing or service cost breakthrough? Three checks might mean that you could lock one or more competitors out of a core market with your new connectivity strategy, locking in customers and making them less likely to churn to another source.

Positively Affect Worker Satisfaction

Does your company have workforce conditions that could potentially affect job satisfaction or your ability to attract or retain talent? Would it enhance employees' desires to work and remain with the company if they had more control over when and where they work, or felt more productive when they were working? Do you want to get younger as an organization? Will new technology create a generation gap? How will you control that? Projects that would improve employee job satisfaction, or help you win and keep critical talent in your company, should get more checks than those that might be hard to win employee support for.

Be Internally Feasible

Take a reality check: What's the degree of difficulty of this endeavor for your firm or group? Do you have the skills and will to pull it off? Be realistic about how much of a stretch your organization can handle. If, in Chapter 8, you scored your firm poorly on its ability to change, you may have to weigh this dimension more heavily.

Be Externally Feasible

Many connectivity opportunities ahead will present privacy, security, and regulatory challenges in the near term, requiring the market at large to adapt. Consumers are deciding whether they're willing to share personal data in more applications; network service providers are only gradually opening up the location information in their networks to outside partners; governments are adding regulations that limit where data can live, what kinds of pricing connectivity services may offer, and more. We talk about the most critical of these issues in the next chapter. Meanwhile, give the opportunity three checks if you don't see any big obstacles, fewer if you see a few or if you could ease the marketplace into this opportunity with smaller phased steps. Give it none at all if there have been recent events related to Anywhere in your sector that could put your firm at risk of a big legal bill in the next year.

Make You Better at What You Do

Beyond winning market share with a more connected product or service, what are the longer-term opportunities for your firm? Jim Collins, author of *Built to Last* and many other books on corporate excellence, researches corporate success over 20-year arcs and longer. That's what you should be considering in this dimension. Would the Anywhere opportunity you've identified build core expertise, knowledge, and insight about fundamentally new, twenty-first–century Anywhere ways to do what you do? All things being equal, improved company efficiency is one of the most important factors to consider. If an Anywhere angle will make your company better at what it does, it should take a lot of negatives for you to have to turn down the opportunity.

If you've taken my advice in this chapter, auditing your opportunities and then assessing their priority along the six dimensions I've proposed, you should have a list of top-flight Anywhere profit ideas with anywhere from none to 18 checkmarks.

Congratulations—now the hard work starts. For your highest-ranked opportunities, you need to determine what connectivity enhancement you would offer, how it would be accomplished, what it would cost in time and expense to do so, and what precisely the expected rewards to your firm should be. You probably already have well-identified processes in place in your firm to develop strategic proposals. The best connectivity vendors will be more than ready to step up to help with that process, providing return-on-investment models, customer references, support for pilot trials, and more. So dig in—but just be sure that the process allows the consideration of the urgency of your situation and the strategic Anywhere benefits you originally identified.

Next Stop: Anywhere

Few firms are where they could be on the path toward profiting from global connectivity. You can start anywhere. Adding connectivity to your products can increase your competitive advantage in the outside market. It can tell you more about your customers and what you can do to make your products more useful to them, driving more revenue to the top line of your company. Adding connectivity to your activities inside the enterprise can collapse time and distance, adding savings to the bottom line, too.

Ideally, you should audit the connectivity opportunities across your enterprise, or your own division or group, and set priorities based on what's measurable, manageable, and away from regulatory murk. Balance the need for short-term benefit with a view to how the best opportunities stretch your organization to be better at its core purpose.

How do you go forward? As with any kind of transformation, there are two extremes at either end of a spectrum of approaches. At one end is what you might call the "Big Bang" method, when organizations spend years and billions of dollars on an endeavor with massive and sometimes ever-increasing scope. Call the other end of business change styles "let a thousand flowers bloom," when companies launch many small efforts, frequently without any overarching guidance, hoping that a few will succeed. Yankee Group analysts guide our clients to chart a course between the two extremes. Many of the executives we speak to in our research command massive organizations, and in their discussions on Anywhere implementation they emphasize the need to be agile and nimble while moving toward a clear vision. See what advice some of them have for you in the sidebar to this chapter (pages 200–201).

* * *

We've encouraged you to ask a lot of questions in this chapter, but here are just three more to consider as you plan your Anywhere profit projects:

1. What connectivity vision will guide what does and doesn't go forward?

2. What criteria does something have to meet to go forward under this mission—how will you define success?

3. What would be the *minimally viable effort*: the definition of the project that would deliver the basic results in the simplest, most direct way, and allow you to build on it later?

Whether you're a member of the C-suite, middle management, or an individual contributor, you can be an evangelist for Anywhere. Advocate for Anywhere by expressing your belief in our globally connected future—including what it can bring us and how fast it's happening. It's this belief, if it takes root in your firm, that will make the biggest difference to your ability to profit from the change. If you're ready to embrace a future of higher profits, lower expenses, and improved productivity and efficiency, then accept the Anywhere commitment, and make connectivity a central component of your business activities. (See Figure 10.4.)

Anywhere Commitment

We will seek to embed connectivity in everything we do—to use the ubiquitous global network to increase our value to our customers, reduce latency in our activities, and limit our consumption to just what we need—because if we do this, we will become a profitable Anywhere Enterprise.

Source: Yankee Group

Figure 10.4 *The Anywhere commitment.*

Chapter 11

Some ANYWHERE Unknowns

"There will be a surprising and chaotic future to this. When powerful forces are unleashed, it's very hard to predict how they play out in combination."

—Dr. Robert Metcalfe, general partner,
Polaris Venture Partners

The Anywhere Revolution isn't something you choose to be a part of; your life and work will be affected no matter who you are and where you are. As a businessperson, the only choices you'll make about Anywhere are in how to take advantage of the changes it brings.

The main forces at work in this revolution are clear, and they are both massive and deeply intertwined. Anywhere is out of our hands because of the convergence of single-purpose networks toward a common digital platform coupled with the rise of digitized media and money, as well as the miniaturization of radio technologies encouraged by the fundamental mobility of the world's population. We can't curb our need to communicate and to do things on the move. The technology is simply chased forward by those appetites.

We've seen in previous chapters just how big the scope of this transformation will be and how rich the opportunities for profit are—in winning new customers and selling them more, by collapsing the costs that hide in time and distance and simplifying our technology platform. You should be well armed with the confidence that this transformation is irrevocably under way and that your opportunities to benefit are real and diverse.

But it's only fair to say that there are some potential issues ahead that could affect the nature of our Anywhere transformation in ways we cannot predict. While these developments wouldn't stop us entirely, they could impede our progress and have a negative impact on our businesses.

We've observed the parallels between the emergence of ubiquitous connectivity and the standardization of electricity. So accustomed are we now to our electrical grid that most of us are completely unaware that the nature of the technology that was ultimately commercialized for municipal expansion was not the same technology that was first unveiled. In a bit of historical irony, while Thomas Edison is credited as one of the giants in developing viable electrical systems (and of that there's no doubt), in fact the technical solution he pioneered was DC, or direct current. By the 1890s he had over a hundred municipal power stations up and running on DC, the response had been extremely positive, and he and his supporters had every reason to believe that it would become the standard across the United States and the world.

But George Westinghouse showed how much safer and more scalable a grid built on Nikola Tesla's alternating current (AC) approach would be. A public and acrimonious battle between the two technologies and their respective champions ensued before AC won and Edison's original DC systems died out.

The point of this story: Most experts at the time would have predicted that DC would win that battle. They would have been correct in predicting the standardization of power across the world; they would have been wrong about which technology would do the job.

It's not only which technology details win that can be tough to predict correctly. As Mike Muller of chip technology company ARM admits, "We can anticipate pretty well what technology will be available in the future; we're less good at predicting what people will do with it." Ben Verwaayen, CEO of Alcatel-Lucent, a leading provider of telecommunications solutions to network operators, agrees: "As an industry we're not a reliable source for knowing what people's behavior will be. We've gotten it wrong many times."

Thus early assessments of the commercial promise of broadband Internet access for consumers were correct—although few experts foresaw the spectacular explosion of YouTube and other media-sharing sites that emerged to use the available bandwidth.

Typically, technology evolutions comprise broad truths that power them, as well as unpredictable developments that complicate, reroute, and confound them. We saw it with electricity and the Internet; it was the same with the printing press and the railway network. Technology transformations since time began are rife with these so-called "black swans," named for the difficulty you would have in predicting the rare, but real, existence of black swans if the only ones you had ever seen were white. The concept has gained in notoriety since Nassim Nicolas Taleb's book *The Black Swan* drew attention to it around the time of the largely unexpected collapse of several storied financial services firms in 2008.

And yet, if we can't know the future, should we do nothing about it? "Uncertainty doesn't mean that you give up, that you stop planning, scheming, and experimenting," says Bob Metcalfe of Polaris Venture Partners. At Yankee Group, we make our best possible assumptions and identify issues that could require course corrections for our clients as they move forward.

So before you finalize your plans for Anywhere, let's take some time here to do two things: first, to clarify a few key assumptions that have powered our thinking; and second, to call out some disruptors that could affect the near-term decisions you make for your business interests. Think of this as the fine print. Forewarned is forearmed.

Key Assumptions: Seeing the Big Picture

To piece together the big picture of our Anywhere future, Yankee Group analysts have pooled a whole range of resources, including hard data and a wealth of insight and experiences shared by industry leaders. Here are four assumptions we've made that are intrinsic to our outlook on the future. We believe we've made the right call, but you should be aware of these as you plan your Anywhere projects.

We've Got the Power

First of all, we assume that the Anywhere Network will be *powered*. We assume that somewhere, somehow, there will be a means of supplying energy to power the Anywhere Network and the devices that use it.

For obvious reasons, this is a reasonable assumption for developed markets. A well-established electrical infrastructure already supports the networks we use today. Energy prices climbing higher will actually drive Anywhere progress in mature markets, as large corporations seek to save money on powering and cooling IT infrastructure by moving services and data to the network cloud.

On the other hand, developing markets in Africa, Asia, Latin America, and the Middle East can't always depend on the electrical grid. They need networks and devices that can run with less power than current developed world models or that can use solar power, diesel fuel, or other more readily available options. Indian mobile network operator Bharti Airtel builds cell sites in some regions using 100 percent diesel fuel. Rajeev Suri, CEO of Nokia Siemens Networks, points out that energy costs in some emerging markets can be 30 to 50 percent of the expense to operate a network. "Operators have to make money serving customers who only spend $2–$3 per month. They will need end-to-end green energy solutions." Some of these solutions will help networks in rural parts of more developed markets. And those rural users themselves need more power solutions to run and recharge their hand sets. Firms like Dutch startup Intivation are working to improve on solar-cell battery technology in hand set applications; recent models provide viable power at a reasonable price for users in the parts of the world with sufficient sunshine.

We'll Pay the Price

We also assume that the Anywhere Network will be *affordable*. Consumers and businesses alike are eager for services and products that make our lives easier, more rewarding, or better in other ways. We're enthusiastic about getting them—but not at any price. And our sense of the price of something isn't just the money we pay. We also have limits to the amount of time and frustration we'll invest in something we acquire. My parents spent one memorable Christmas Eve assembling a bicycle as a present for the next morning. Struggling late into the night, frazzled and bleary-eyed, they finally found a missing piece of the instructions—including this priceless warning: "Assembry [sic] of bicycle require great peace of mind."

Some Anywhere ideas have already floundered because of poor pricing. Network operators hope that consumers will sign up for new services on their networks, beyond voice and basic data transmission. In 2008, AT&T launched a Web-based video monitoring application (transmitting video from Webcams in the home to the Web, controllable through a browser) called the AT&T Home Monitoring Service, for which it charged $9.95 per month in service premiums. While U.S. consumers confirm interest in such expanded services, with an average monthly household spend of $42 for broadband and $67 on mobile, they have proved resistant to buying new services when it means a substantial ongoing addition to their monthly spend. AT&T suspended the service offering in 2009.

In Europe, mobile network operators spent huge sums of money to upgrade their networks in the 1990s to be able to support data transfer at higher speeds—premising those investments on assumptions of pricing for the eventual experience that, when launched, ultimately left consumers cold. It wasn't until 2008 when operators, anxious to win mobile data users after years of disappointing uptake, began offering the service at a flat monthly rate rather than a per-bit charge that demand exploded. Yankee Group analyst Declan Lonergan explains: "With pay-per-megabyte pricing, consumers have no idea how much they're spending every time they press 'browse' on their phones, so only the bravest souls do so. Flat-rate pricing for mobile broadband is the obvious winner; it meets the consumers' criteria for simple, intuitive, good value, and obvious spending control. The most compelling evidence of its success is that all network operators have adopted it, and consumers continue to flock to the offers."

In the enterprise arena, early telepresence systems proved far too expensive for most firms, at upwards of $600,000 for a two-location system. While the experience offered was dramatically better than current videoconferencing, initially the cost-benefit equation didn't compute. As costs come down and the desire to reduce travel expenses increases, the equation for more firms is gradually tipping in favor of telepresence. In 2009, Cisco reported that it had sold over 2,300 units to more than 350 customers around the world.

The price of getting onto the Anywhere Network in emerging markets comes down to what access costs a consumer there—whether it's a computer, a mobile phone, a netbook, or a telecenter providing the on-ramp. Even though retail prices of mobile hand sets in those markets are continuing to fall, the price is still a substantial obstacle, if not downright impossible, for people of limited means to meet. U.K. startup Movirtu came up with a creative system it calls "share-paid" to let multiple people share a mobile phone, but each with his or her own phone number. CEO Nigel Waller reports that the first systems will start in sub-Saharan Africa in 2010. "A new phone with a warranty in those markets can now be had for about $25; that's inexpensive but still out of reach for many. Of the 3.5 billion people today who don't have a phone, about 1 billion of them borrow a friend's or pay to use one at a kiosk. We can give them an account of their own, which they can top up and even get loyalty points for. For the person sharing the phone with them, they can have a share of our revenue, essentially a 'thank you' for hosting the caller's activity."

Pricing built on supplier-centric assumptions may fail. Your job in building well-priced Anywhere Experiences is to understand how the users view the total cost compared to value received and how this equation will differ significantly in the varieties of markets you serve.

The right price is going to be crucial for everything from inexpensive mobile hand sets for the poorest of tomorrow's Anywhere Consumers to very clear ROI models for deploying unified asset visibility programs in large enterprises.

Why do we assume that Anywhere Experiences will be sufficiently well priced to support our demand outlook? Simply put, because pricing is something that's easy to change. Besides the constant reductions in cost delivered by increasing sales volume, history shows that motivated suppliers of all kinds will experiment with mixes of value and price until the market finds a combination that it likes.

But power and price aren't all that it will take; there are two more assumptions intrinsic to the Anywhere Network's expansion.

Open for Business

We assume that the Anywhere Network will be *open*.

The main explosion of external value added to the Internet in the 1990s came from its simple and widely promoted standards for connecting to the Web: HTML and the browser interface. It's a large part of what has since led to Facebook and other businesses building billions of dollars of value on top of the global network. Because the standard for creating files viewable by a browser was open and easy to use, and because those Web sites were reachable to anyone with a browser who could type in the site address, use of the Internet grew exponentially.

There are two separate and equally important aspects to open access that we assume will characterize the Anywhere Network. The first is that the network is open to content and service providers so that they can offer their services to anyone. The second is that consumers of content and services have ready access to content and that they're not blocked from getting to it.

In 2006, two popular social networking sites were launched in the United States and began to compete for users both in North America and abroad. That year, MySpace boasted 75 million users; Facebook about 12 million. Then in May of 2007, Facebook published its application interface, making it possible for virtually anyone to build an add-on experience for the site. In less than a year, its site's basic functionality, developed and maintained by Facebook, was further embellished by over 7,000 applications developed for it by the crowd—thousands of users and businesses who took its published specifications and used them to create games, educational tools, self-help aids, and more, all on top of the Facebook experience. Facebook usage soared, making it the most popular online community in the world; Web monitoring site Alexa estimates that over 20 percent of the world's Internet users visit it each day, averaging almost 30 minutes per visit.

If you've grown up with the Internet, perhaps this feels obvious—the way the world *should* work. But it's a distinct shift in mindset from twentieth-century single-purpose networks. Telephone networks were completely closed systems, controlled from end to end by their operators. Phones were rented to users by the phone company; the big news in the 1950s was the introduction of hand sets in a color other than black. Until 1968 in the United States, it was actually illegal to connect any item that was not the property of the phone company to the AT&T network. Following a change in regulations, a market for tremendous new variety in telephone hand sets, answering machines, fax machines and more emerged.

Open networks are those that allow users access to what they want and need over the network, as opposed to closed networks that limit access to network-based services to just those that their network service provider thinks they need. As a result, open networks beget expanded use.

Open networks also allow external developers of network-based services to get access to data and intelligence inside the network, such as a user's location, which they can then use to develop new services on the network. In Europe, digital map supplier Navteq (owned by Nokia) enables advertisers to reach mobile consumers with their brand message by leveraging the map data in their location-aware device. "Building around an advertising-based business model can open up additional opportunities to developers, as another way to generate revenue from their own creativity," says Chris Rothey, vice president of market development and advertising with Navteq.

The "long tail" concept—a term first coined by Chris Anderson and detailed in his book *The Long Tail*—describes the phenomenon of an inexpensive network allowing the wildly diverse interests of small groups of people, physically distant, to be economically served. The incredibly diverse appetites of a complex market need the power of the market itself, willing and eager to use tools to build things to meet its own appetites. The launch of the Apple iPhone App Store, allowing the market to propose and deliver applications for the iPhone with Apple as the distributor, brought over 50,000 applications to the world in just the first 12 months of operation. While in many respects the iPhone is not a very open device nor Apple a very open enterprise, Apple has published key specifications and information that external application developers need to create software that runs on the unit, and is serving as the distribution solution for those same developers, allowing them to reach a sizable market they would be unlikely to be able to do on their own. (See Figure 11.1.)

To expose the tremendous potential expansion in connected devices in the world, network operators need to make it easy for consumer electronics manufacturers and others to build devices that successfully connect to the network and exchange data. Sprint/Nextel CEO Dan Hesse says, "There will be new creativity in devices that we

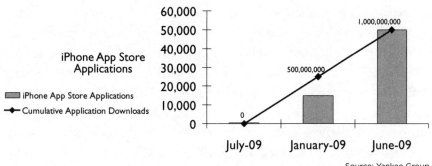

Source: Yankee Group

Figure 11.1 *iPhone application growth.*

don't see today. And network intelligence is every bit as important as network capacity. Openness is more of an opportunity than a threat. We opened our phones to full HTML browsing early, so consumers can go anywhere they want."

Yankee Group analysts assume that for widespread adoption and opportunity, our wireless and wireline broadband future will be based on networks that are open both to consumers able to navigate them freely and to third-party service providers able to build value-added offerings on top. While technology's history books include many episodes of closed technology solutions tightly controlled by suppliers—and even today, Apple's portable devices don't allow owner battery replacement—the massive explosion of the Internet was based on open standards, creating new widespread assumptions in the marketplace of users that will be nearly impossible to overwrite.

As you plan Anywhere products and services, you'll want to consider your network operator partnerships with care. Understand their plans to be open; compare their words and deeds with those of their competitors. You need to think big, not exclusive, choosing the network partners with the best ability to attract crowd-sourcing to their networks and to whatever solutions you build atop them as well.

Adopt This Experience

Finally, for the revolution to continue as we predict, we assume that Anywhere Experiences will be *adoptable*. We assume that user experiences with new connected products and network-based services will have the ability to be referenced and will be repeatable so that other firms adopt them where they suit and so that users spread adoption by word of mouth.

In the early days of the Internet, many of us were given pause when Web sites greeted us with lengthy forms to fill out, asking for personal details like our address, income, and more before they shared content or commerce opportunities with us. The commitment of time, as well as the unexpected sense of personal invasion, caused many consumers to click away to other sites, looking for less invasive options. It took a while for businesses to adjust their requests for information to balance the perception of value we had as consumers to what they offered us.

But for firms unable to recapture the full cost of their Anywhere offerings through the cash price users pay, the best alternative may be to subsidize the users' experience through advertising revenues. Will we trade a willingness to receive marketing messages in the form of ads or coupons on our mobile phone for goods and services to get a lower-cost phone service? It will take time for that to evolve in phases of trial and error.

These offerings will be about information at their core, and that information —our shopping preferences about brands, stores, and payment methods, for example—will depend on our being willing to trade more information about us and our environments to get the core value from the offering and to feel comfortable that the information will not be misused.

We assume that this adoption will take place, because consumers will gradually accept the value exchange and because the services and experiences that will evolve can be refined. Unlike massively complex products that take years to design and build, such as airplanes or cruise ships, these are products that can be publicly trialed, revised in a matter of days if not hours, and offered again in a relatively simple and potentially infinite process of trial and error. The vast majority of new Anywhere Experiences will be presented to the broad public in the open light of day, allowing for a rich interchange of experimentation, adaptation, and refinement.

How risky is it to assume that the Anywhere future we are moving toward will be powered, appropriately priced, open, and sufficiently "refinable" to be widely adopted? Not very. While we're unlikely to successfully predict the fate of every individual firm or specific new Anywhere service, these basic assumptions are part of the broad context for your efforts.

Primary Disruptors to Your Anywhere Profit Opportunities

So much for the assumptions we've made in our consideration of the future of Anywhere that we believe are reasonable to make. Now let's consider what monkey wrenches could come your way as you identify the best opportunities to simplify or augment what you do with expanding connectivity.

Remember that black swans, whether good or bad, are named that for a reason. So in providing you with some cautions about future unknowns, what you won't find is a complete list of every possible nasty surprise that may befall us. But the most disruptive events may come in four principal areas—the provision of sufficient network capacity, the risks to privacy and security, the intervention of governments, and the inertia of the past. In each arena, you'll need to stay alert to potential threats and to develop strategies that navigate around the worst risks.

Bandwidth Bottlenecks

Anywhere Experiences will demand capacity in the global network to move bits around. Global networks will have to continue to add capacity in two places: first at the edge, where we as users connect to the network from our home, our wireless device, or some other access point, and second, in the networks' core innards where our digital activities are bundled for bulk transmission like so many containers in an ocean-going cargo ship.

That requirement of networks to grow capacity both at their edges where users transmit and receive data, and in their cores where bundled traffic is shipped between midpoints on higher-capacity links, has been growing exponentially for some time. In the spring of 2006, Google was performing about 100 million searches per day for its users; in the fall of the same year, it purchased video sharing site YouTube, which reported at the time that it was delivering about 100 million video streams per day. Just three years later, Google's reported searches per day had grown by a factor of three, and YouTube confirmed that it was streaming over 1 billion video clips per day, or a factor of 10 in the same three years. Demand to use the Internet for all kinds of activities has grown explosively in the last decade, with a concomitant impact on network load.

And it will continue to do so for the next decade as well. Consider the billions of users and trillions of devices we're anticipating to be Anywhere. These will result in, literally, quadrillions of packets of bits moving around the world, delivering our photos, movies, device readings, navigational coordinates, and oh yes, our voices. Those are the bits required by the expansion of activities we can anticipate *today*. What about Four-Dimensional Smell-o-Vision? No such thing exists today, but that's my own personal placeholder product concept of the future: some weird yet wonderful black swan experience that we haven't yet imagined. We should assume that many more YouTube and SecondLife concepts will emerge, all placing additional load on the network.

In theory, the network's capacity could expand almost indefinitely. And in the past, broadly, it has responded to the demand with appropriate supply. But can that continue? Disruption could be caused by several things, including the following.

Delays in Wireless Spectrum Availability

Delays or inefficiencies in making new radio spectrum available could slow down Anywhere. Governments around the world control the radio waves and license them to others for use. Since regulation is often a slow-moving process, it can average several years between the time that a government decides to move to the time that the winning licensee actually puts the spectrum to use in creating more capacity for its customers. The longer the process takes, the more it will be delayed by administrative shifts and a lack of consistent direction from governments, which often causes the private sector to hesitate in making its own plans. Reed Hundt, chairman of the U.S. Federal Communications Commission during a period of tremendous change in communications regulation between 1993 and 1997, says that he currently sees several worrisome trends in wireless spectrum allocation around the world. "In India, the spectrum slices they're auctioning off are inefficiently small. In China, their choices of spectrum to license are coupled with technology choices that represent [companies that are] national champions being pushed forward, instead of a commitment to the economics of scale provided by using global standards."

Given the challenges that governments the world over have in matching their funds to their expenses, the potential fees gained by licensing firms to use that spectrum have tempted them in the past into short-term policies that emphasized near-term revenue generation for the government. Sometimes this came at the expense of the viability of the operators' business models. However, the main issues currently inhibiting effective spectrum use are less about the approach governments have to charging for them, and more related to freeing up encumbered spectrum, expediting auctions, and lifting use restrictions.

For a taste of how regulation could affect the growth of the Anywhere Network, see the sidebar (pages 230–231) on the U.S. start-up M2Z Networks, Inc. Your planning for Anywhere products and services needs to be sufficiently agile to take advantage of the emergence of disruptive competitors like M2Z should they be able to get their plans off the ground.

Wireless Bandwidth Expansion Investment

Anywhere Experiences will depend on wireless networks. We're a mobile species, after all, and we want things with us as we move around. This desire is at the core of the Anywhere Revolution, and air as we've said is cheaper than fiber or copper. Unfortunately, radio spectrum—the waves that help transmit Anywhere Experiences from one person or thing to another—is not limitless. Those waves also have owners who decide how, when, and why they get used.

One problem could be delays in upgrades from mobile broadband to more advanced technologies. The transfer from current mobile broadband networks, called third-generation technology, or "**3G**," to "**4G**" or fourth-generation solutions, is going to be costly; Yankee Group analyst Phil Marshall estimates the incremental cost to the network operator on a per-user basis will be anywhere from $50 to $150 in capital expense after mass-market scale is achieved. Mobile networks will meet the growing appetites for data transfer, but it's not going to happen quickly, it won't happen completely evenly around the world, and, to quote an old English proverb about things going wrong, there could be many a slip 'twixt the cup and the lip. In Figure 11.2, you'll see Yankee Group's current outlook for the global number of mobile subscribers by region connecting via advanced 3G and 4G broadband technologies.

If they don't expand, the satisfaction levels of users for bandwidth-heavy mobile experiences like video will be poor. This will impair

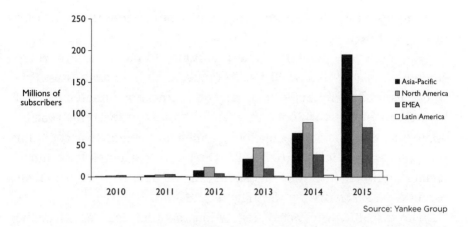

Source: Yankee Group

Figure 11.2 *Forecast for growth in 4G network subscribers.*

consumer willingness to adopt them or to pay for them, and it could certainly delay the emergence of Anywhere by region or even by critical market segments such as Actualized Anywheres.

Next-Generation Network Architectures
It's not just radio waves that need to carry Anywhere bits around the world; much of the existing global network infrastructure is composed of cables: glass and copper wires that traverse our oceans and neighborhoods. The exponentially growing demand we have seen and which will continue affects how crowded those highways are too. History has proven that, over the long term, networks increase in capacity to match growing demand, but this capacity expansion isn't equally distributed: There will be many localities where the existing network infrastructure is insufficient, thus disrupting the Anywhere products and services you may be creating.

The communications industry uses the term "next-generation network" to describe the transition that's under way to an all-IP network that carries any of its information around from place to place in IP packets, as we discuss in Chapter 2. For operators of existing networks around the world—including many former monopoly telephone service providers, cable TV service operators, and others—this transition requires at least three things: massive financial investments, regulatory clarity, and a long-term approach to calculating the return on their investment.

Will network operators spend the money to build the Anywhere Network capacity that we'll need? Yankee Group analyst Brian Partridge estimated in 2009 that telecommunications-related capital investment by network operators would drop over the prior year by approximately 4 percent, as operators reined in spending to account for the poor economic environment. "Operators are employing much greater caution in how they invest their capital budgets relative to their revenues," says Partridge. (See Figure 11.3.)

The disruptive impact of regulation rears its head again in the planned transformation of existing fixed-line networks to next-generation networks. As governments try variously to protect the interests of incumbent networks, encourage more competition in network provision, and ensure the build-out of networks to reach all their citizens, regulation emerges that network operators fear could limit their ability to earn a return on their financial investments. In the United Kingdom, for instance, the British regulator Ofcom and incumbent network operator British Telecom (BT) engaged in a complex ballet surrounding BT's willingness to invest in upgrading its network and Ofcom's goal to

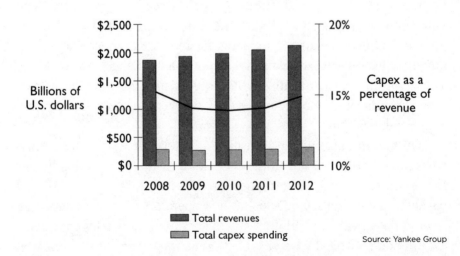

Figure 11.3 *Global network operator capex forecast.*

ensure that parts of that network can be made available to BT's competitors. This issue is playing out in similar ways around the world in both developed and developing nations where regulators are focused on the upgrade of networks operated by incumbents.

To recap: networks grow, and governments regulate—but both do so in unpredictable ways. Will the Anywhere Network have the capacity your Anywhere products and service offerings need to satisfy the demands of your customers? Many years ago, I was marketing communications equipment and struggling to understand the needs of a particular customer. "It's simple," he sighed. "I just want the speed of light for pennies."

Sounds good. While we all might want it, no one promises it will be there at that price when we need it. You should not attempt to launch major new Anywhere offerings in any market without investing the time first to understand the demands you may place on the network, the capacity of various network providers to service those demands, and the potential of regulation and pricing that may affect the viability of your plans. For bandwidth-intensive offerings, consider tiered experiences that adapt the quality of experience to the capacity available. Regulatory monitoring should accompany any long-term strategy you put into place to expand your geographic reach.

When Bad PR Is Just Bad PR: Possible Anywhere Backlashes

If it's true that technology waves are dependent on a virtuous circle of adoption, then large-scale advances cannot be forced. They have to be accepted by society, by individuals, and by key groups and organizations. To be clear, not every concept or offering associated with the emergence of ubiquitous connectivity will succeed, and for those that do, their adoption could still come at the end of a bumpy road. The risks around adoption of Anywhere Experiences present our second area of potential black swans to consider.

There will certainly be many Anywhere offerings and experiences that we will reject. Sometimes products are mismatched for their markets, such as the Zune—aimed at teens and 20-somethings in love with the iPod, but with the complexity of features only accepted by older, more technically inclined buyers. Others we won't appreciate, as with my electric knife, lying forgotten in a cabinet somewhere. There will also be disparity of adoption from one generation to another; our children will adopt more Anywhere technologies than we do. All these aspects of adoption are to be expected, and we're not concerned about them in our list of potential disruptions.

But what could substantially derail your plans for Anywhere success would be any kind of disastrous or horrific event involving consumer safety associated with a ubiquitous network. You probably know the saying, "There's no such thing as bad PR." But what if a child is kidnapped by someone who was able to track her walk home from school using mobile phone location technology? What if an elderly couple's in-home health monitoring is seriously disrupted by an unplanned network outage? What if faulty RFID technology certifies food as safe that ultimately sickens its consumers? These and many other tragedies could befall individuals, as tragedies always do. Airplane crashes, credit card and identify theft, accidents that hurt or kill people are a fact of life. In the commercial arena, misfortunes associated with new technologies gain the media's attention and can put the brakes on the population's adoption of related experiences.

Any technology can be used for good or bad. But if a child is harmed through the misuse of personal information transmitted through the wireless networks, the public outrage and media frenzy could delay the release of other viable useful services, even as the network becomes the police's best weapon to catch the perpetrators.

As business leaders, we have to be thoughtful about how to introduce new connected experiences. We will need to monitor related offerings for lessons we can incorporate into our own systems to ensure that we gradually bring our consumers along with us, but only as they're ready to take further steps in sharing information and trying new things. However, we can be as thoughtful as we like, and consumers can still take us out of our comfort zone.

How can we navigate consumer trust issues and other risk factors? You will need to reassure and communicate with customers about the use of their information. The ease of sending bits Anywhere makes mistakes as easy as success. Amazon stumbled in 2009 when it used its persistent connections to Kindle owners to summarily delete purchased digital copies of material that it learned it hadn't had the right to sell. Device owners howled, and the firm conceded it had acted in haste. Expanded digital connections to customers come with added obligations.

It will be more important than ever to stay abreast of legislative and regulatory developments in your markets that touch on consumer protection. The European Union has powerful consumer protection legislation concerning data privacy, which ensures, among other things, that European consumers have a right to be notified when data about them may be collected, to know what the data are going to be used for, and to withhold their consent. If your offering hews to European Union directives, your business interests will not only be well protected, but it's also more likely that you will have a strong base from which to build trust with your customers.

Avoid concepts that depend on centralizing mass amounts of consumer data. Instead think about ways to federate available data across multiple sources using the network. While in many areas, such as health care, the economics of scale and the potential for consolidation of data suggest just that approach, the rights of users to privacy may supersede those approaches. "Do you think it's politically feasible to create a national medical database? Is it workable from a privacy perspective?" asks CareGroup CIO John Halamka. "There won't be a giant database in the basement of the White House; that won't happen. The best you can do is to enable various databases to be queried by Web-based services that depend on the patient's permission."

"I'm from the Government, and I'm Here to Help"

Global economic recession is causing governments to look for ways to restart growth in promising sectors. Yankee Group regulatory expert Dianne Northfield points to stimulus commitments around the world totaling in the trillions of dollars that are intended to expand network infrastructure in countries including Malaysia, Ireland, Portugal, Greece, the UAE, and South Africa. In the United States, federal stimulus legislation passed in 2009 set aside billions of dollars to help fund the expansion of broadband networks in rural and urban areas.

But the movie stereotype of the busybody government official has a foundation in reality. Sometimes government help comes with strings that can constrain innovation. Bob Metcalfe describes the context of government activities: "Governments, well meaning as they usually are, tend to support the status quo. They end up making rules that slow down the spread of innovation because they're protecting existing investments, sunk costs, revenues." Paul Sagan, president and CEO of application and content delivery provider Akamai, worries about how helpful recent government initiatives will be. "We need to find ways to encourage the fastest possible growth of broadband, particularly in the United States. We need to think about what the government can do and to be very wary of what it shouldn't do to retard growth."

Regulatory Uncertainty and Politically Motivated Restraint
Recent events demonstrate how governments and underlying policy issues can be a problem for the expansion of Anywhere. In 2003, the calling service Skype was introduced to the mass market, creating a revolutionary alternative for making long-distance phone calls using the Internet. Phone calls that previously could cost hundreds of dollars using traditional phone networks now could be made for pennies. How Skype was treated in various jurisdictions around the world was in part due to the challenge it presented to incumbent network providers and traditional concepts in communications regulation. In China, ambiguity around how to classify Skype's service led to no established regulations either for or against it. Providers have emerged, but reports have surfaced of incumbent network providers blocking access to their services. Operators are still blocking Skype-type calls in many countries around the world; the European Commission has been investigating and is likely to pressure operators in the EU to remove such restrictions.

Governments concerned about controlling political unrest continue to try to block dissemination of information and the organizing of opposition using the global network; a few recent examples included presidential elections in Iran and the anniversary of protests in Tiananmen Square in China. Although it has yet to happen on a large scale, if governments successfully lock off parts of the global network, it could impair our opportunity to reach new consumers with commercially valid offerings.

Rights and Liabilities

Intellectual property and legal liability are a major issue as well. As information travels more freely around the world and networks carry ever more varieties of activities, laws struggle to keep up with the rights and responsibilities of people in the process.

The European Union, for example, is crafting new rules intended to harmonize and extend its member nations' approach to laws about network traffic. If its final legislation requires Internet service providers to actively monitor traffic to ensure the legality of its contents, it could add expense and huge financial risk to networks, and this could have a braking effect on the expansion of networks serving EU citizens.

Net Neutrality

A major source of contention surrounding some governments' intervention in the growth of the global network infrastructure has been labeled "network neutrality." The term refers to a principle that network providers should carry traffic indiscriminately, for instance, not blocking or slowing contents of some particular type (e.g., video-sharing traffic) or source (e.g., a service from a company a network operator might regard as competitive). In the United States, Canada, and elsewhere, legislation or regulations have been proposed that aim to ensure that network providers aren't legally allowed to do so. Proponents of these approaches hope to ensure that the digital network is not unfairly compromised by its carriers for their own profit incentive. Critics of the legislation worry that inappropriate restraints on what network operators are allowed to do in order to maximize the commercial returns on their investment create unfair competitive advantages for other services over their own. There could also be negative consequences for consumers, such as poorer network performance for everyone as a result of the actions of a few extremely heavy users.

As with many impassioned debates, there are legitimate concerns on both sides of the argument. In a completely "net neutral" world, it could be that networks won't continue to upgrade supply to match demand because they could lack the confidence that expanded capacity could be rewarded by expanded revenues. In a nonneutral context, companies like Google that want to offer Anywhere services on top of the network, such as movies on demand, could potentially be prevented from getting their video streams to a consumer if the consumer's local service provider decides to slow, or lock out, that traffic to try to favor the consumer's purchase of similar programming from the network provider itself.

Not everyone in the sector would agree with this perspective, but at Yankee Group, our view is that legislating the ways in which network capacity can or cannot be monetized by the networks' owners is likely to stifle their efforts to find better ways to monetize the constantly increasing traffic loading their networks. We don't support formal legislation, but we do believe that there should be some principle ensuring nondiscriminatory carriage of like content.

How can you navigate the uncertain governmental context for Anywhere that's still ahead? Remember that this is a revolution, and as such it will shift the commercial power structures currently in place. In any transformation, regulation lags innovation by decades or more. By now most governments recognize the value-creation potential of digital connectivity for their citizens. As a result, government-led initiatives may be intended to encourage the continued expansion of the network, but they may still complicate the competitive landscape over the next decade, particularly for network operators.

Prioritizing markets for your Anywhere products and services will require not just an awareness of the state of a region's broadband infrastructure—its Anywhere Index—but also the relative clarity or uncertainty of the legal infrastructure for the products or services you hope to offer. Matt Dill of Western Union, in talking about the challenges his firm faces in expanding its mobile payments and remittance services, says, "The relevant regulation today is oriented to banks and physical cash retail activities. Just because you can do something operationally, may not make it workable on the regulatory side. We may need a bank to provide regulatory integrity to our offering in a country. The ability to be flexible as you go around the world is as important as the need for a vision to make it work."

The Legacy Coefficient of Drag

A famous remark by retired business executive Jack Welch goes something like this: "There are some people whose fondest hope about tomorrow is that it looks just like yesterday." There are firms and leaders for whom the Anywhere Revolution brings such fundamental disruption to their value proposition that their principal energies will inhibit its emergence.

Incumbent network operators offer a great source of new Anywhere experiences going forward. As important partners for many Anywhere Enterprises, they can help us virtualize our businesses, move computing to the cloud, and reach new users for our services—the big three. But many aspects of Anywhere threaten their future as well.

The opportunity for Anywhere is collapsing traditional industry boundaries; incompatible business models collide, and overlapping aspirations among ecosystem players disrupt the status quo. Network service providers no longer have the exclusive claim on value creation, but rather are under siege from players throughout the entire ecosystem scrambling to expand their shares of network users' wallets.

As the Anywhere Network continues to unfold, revenue opportunities will splinter among a plethora of services and applications that traverse communications, the Internet, media, mobility, and machines. Historical business models based on "owning' a customer are going to change. At the least, network operators are going to be very worried about being forgettable to their customers. How many people today can actually name their energy provider?

Trying to preserve the historical business model is going to be tough for network providers. Many were or still are monopolies or virtual monopolies for telephone or TV service. As one common network can do all those things, each of the network providers is looking at several new sources of competition. We could see delays or disruptions coming from network providers attempting to use government regulations to delay the competition from Internet service providers.

Operators of existing networks are probably the most easily identified source of drag on network change. But we should rise above the

Any Anywhere Network Black Swan

While our progression toward a fully connected future is in no doubt, there are numerous issues that could color the way in which this connectivity emerges. "Black swans" are the unexpected developments—positive or negative—that could have a dramatic impact on your plans to profit from Anywhere. One such possibility is a free wireless broadband network that entrepreneurs John Muleta and Milo Medin, cofounders of start-up M2Z Networks, would like to launch in the United States.

Here they talk about what they want to do and why, and they identify from their perspective some of the challenges to their plans. As you read, consider this: Whether it's M2Z or another supplier, whether it's free or just priced very differently from today's options, it could happen. How will you prepare your business for the emergence of a game-changing alternative to today's broadband landscape in the markets you serve?

Q. *What is M2Z Networks trying to do?*

Medin: I was with one of the earliest cable broadband Internet services in the U.S. In 1996, we charged $40 per month to provide 4.5 mbps of download capacity to the consumer's home. Twelve years later in that same area, consumers pay $40 per month and get 6 mbps. But in that same time frame, the price of long-distance phone calls came down to pennies a minute. In that same time period, cell phone calls that used to cost $1 per minute now cost around 5 cents per minute. Laptops that used to cost $3,000 now cost $300. So why hasn't consumer broadband in the United States gone through that same price erosion curve?

We see an opportunity for a disruptive competitor to enter the scene, to try to move consumer broadband pricing to where the rest of the consumer's technology-based experiences have gone. And if you want to force that price erosion, then the medium you have to look at is wireless networks, or radio spectrum. It is the second half of the broadband revolution.

Q. *You want to make the broadband network experience free to consumers. Why?*

Medin: I was raised on a farm with no cable TV, no fancy stuff. Even so, I wasn't too disadvantaged compared to my friends who did have those things at that time. But today, giving a kid a computer without broadband is kind of like giving him a textbook he can't take home from school. Broadband is a huge competitive differentiator now.

Muleta: The adoption rate of broadband service in the United States today is skewed to income. If you have a lower income, you're rural, or ethnic just joining and learning America's economic model, you're slower to adopt services, especially if there's a significant barrier to getting it in the first place. But there is exponential adoption for services that have a free component. What it says is that affordability is a big barrier to purchase and to try.

The driver behind free services is not just targeting whoever can't afford the for-pay model. It also benefits everyone else who wants to be able to reach those people. Banks, utility companies, anyone who sells to consumers broadly, they all want universal connectivity.

In some ways, people online want the ones who aren't online even more than the off-line people want to get online themselves.

Q. *How would your free network work; what would it feel like?*

Muleta: We expect to have two kinds of devices for consumers to access our wireless broadband network. One will look like the routers that are already in consumer homes to provide Wi-Fi access to a network connection. Consumers will be able to go to any store, buy a box, turn it on, and be connected to a broadband network instantly. No setup, no waiting for an appointment. It will be just like buying an appliance.

There will also be laptops with built-in M2Z transmitters. This is because computer manufacturers have shown interest in doing this. Imagine what it would be like for them to ship a computer to someone's house and know that it's connected. When there's a problem, they could have a connection to that computer almost right away.

Q. *When do you expect be in business?*

Muleta: The timeline for when we can do anything depends on the Federal Communications Commission (FCC). We're advocating for an auction of the spectrum we would use to do this, to let the market speak. The auction can show the government who in the market wants to provide services using this resource. Eighteen to twenty-four months after the spectrum is made available, we could have the first parts of our network up and running.

Medin: Worst case, though, the timeline for M2Z Networks' rollout could be never. The spectrum needed for this is available, and it's ready to go. But existing providers would prefer to bury it. If that happens, it's going to take a long time. The FCC doesn't have any other spectrum in the queue to auction that you could run a broadband service with. If the commission wants to make more available, it could, but the average time frame is seven years after new radio spectrum is identified to turn it into services for the consumer. No fight is ever won until the spectrum license is assigned. We are facing large, well-funded incumbents that don't want this to happen.

noise of the net-neutrality debate to recognize the other potential obstacles to the openness we advocate in our Anywhere vision; any closed solution that doesn't promote sharing of content between users, and interoperability between the different devices used by users, impairs this picture. Sadly, this describes ever-popular Apple.

Many other sectors of the global economy are still dependent on "somewhere" businesses and consumers. The Anywhere Revolution may not be welcomed by the automotive industry, the commercial real estate market, and other sectors that depend on the current model of work based on collecting people and assets together on a more permanent basis in specific locations. Disruption could come as these industry players try to use whatever means they have to prevent changes they see as harmful to their interests.

But compared to the delays or disruption that could jeopardize your success with Anywhere initiatives, the potential drag within your own firm may be at least as significant. In Chapters 9 and 10, we identify your need to assess your enterprise's ability to change and to factor that into your prioritization of Anywhere profit projects. Michiel Boreel is the CTO for Sogeti, a worldwide technology services firm. "The speed of innovation is not dependent on technology; it's dependent on leadership. If leaders have a reason to move faster, or if they believe in the power, or they're faced with big problems they don't know how else to solve, then they move." Can you frame your Anywhere opportunities as a solution that addresses a critical business imperative for your firm's leadership?

We've looked at the pace of Anywhere around the world, noting that Anywhere opportunities will be pegged to the network's expansion. And we've pointed out that the readiness of your firm to tackle the opportunities being created will be critical to your success. But Anywhere profit will also be dependent on a careful study of potential disruptors.

In our final chapter, we talk about some of the larger implications of the Anywhere Revolution. I have been tremendously inspired at what the digital network has already contributed to our lives and what it is likely to do in the future. I think it can do the same for you. Before you set off to profit from Anywhere, that will be my final gift to you.

ANYWHERE: Final Thoughts

"The real meaning of ubiquitous connectivity is what it will mean to society as a whole to not have an isolated class."

—Milo Medin, cofounder, chairman, and CTO, M2Z Networks

A short but memorable film made in 1977 by Charles and Ray Eames called *Powers of Ten* has engrossed millions of people since its release. It travels from the outer limits of space, through the hand of a human, down to a single proton—zooming through a total distance equivalent to 40 orders of magnitude in just 9 quick minutes. Today, Google Earth lets us zoom in and out of our world in seconds. How long will it be until Google—whose jaw-dropping mission is to organize the world's information—augments that experience with the ability to zoom all the way to outer space and back, documented with photos, landmarks, and layers of other information?

This book has examined a set of technology changes to see how you can profit from them as a businessperson. We've looked at the factors leading to the emergence of ubiquitous connectivity—the development of a common network, the appetite for broadband experiences, and the compelling economics of wireless communications. Their convergence into a powerful, pervasive digital Web helps us scratch the very deep itch we all have to take our experiences with us wherever we go. And it allows some of us to participate in the global economy for the first time.

We've also seen what connectivity, when married with a company's assets and activities, can do to reduce time and distance and to limit the complexity of our infrastructure so that we can concentrate on what we do best as a business.

In all those developments lie chances for businesses to capitalize handsomely: selling us new things, using these expanding pathways to reach us with more value, and bringing information to the point of need with the speed of light.

I won't apologize for the profit focus; I'm a businessperson too. But given the opportunity that Anywhere presents for enlightened self-interest—doing things for others to ultimately help ourselves—I think we should use our last chapter together to take a step back: to zoom out from the revenue and cost discussions here to consider some of the larger implications of the world's move to Anywhere. Then we'll zoom back in right to you as a businessperson so that I can leave you with a final few words of advice.

The Anywhere Big Picture

We've used lots of big numbers in talking about the scale of the coming impact of Anywhere—billions of people, trillions of devices, quadrillions of packets flying around the world. Given that, it would be pretty surprising if Anywhere didn't also bring some large implications in our world beyond the commercial opportunities this book has considered. Our Anywhere future will be dramatically different in economic, social, even moral terms.

The Anywhere Economy

Clearly, the profit potential of Anywhere is exciting; we've spent the better part of this book exploring it. In many respects, Anywhere is all about the money, since ubiquitous connectivity will make it easier for us to share in the world's riches. Behind the profit itself, though, lies an interesting shift in economic power.

Anywhere brings immediate information and options to everyone. In economic terms, it shifts power from the supply side of the market to the demand side. "Connectivity is changing our world from a supply, or seller's, market to a demand, or buyer's, market," observes Paul Sagan with Akamai. "In the past, sellers had more power. You decided what hours you were open or closed, and you could see your competition by driving up and down the street. You could force your market to get your brand messages. Customers had very little alternative messaging and not a lot of independent reviews for products."

In the Anywhere Economy, customers take control. It began on the Internet in the mid-1990s, as car buyers began to go online to learn the true cost of a new car and arrived at a dealership better informed than the dealer's own salespeople.

The Anywhere buyer is empowered both by knowledge and by choice. We all know now to price-shop online, and as connectivity expands, more vendors around the world will offer their products to buyers all over the world. As they do so, the pressure on prices won't come only from the increase in the number of smart customers, but from new Anywhere Enterprise competitors as well. If you're not using every possible digital pathway to a customer, you can bet someone somewhere will be doing it in your stead.

The phenomenon spans all sectors of the global economy. "Health care was doctor-centric. With ubiquitous connectivity, it will become patient-centric," acknowledges CareGroup's John Halamka. Happily more citizens will get good health care because access to it will be much less dependent on location. Medical experts on the other side of the world will review X-rays and other medical data from a pregnant woman in Africa, applying their expertise to diagnose risk factors that less-experienced local staff might have never seen. An expert surgeon will be able to oversee an operation even when it's not possible for him or her to be physically present.

Knowledge and reach will emerge as the key currencies in the Anywhere buyer's economy. Success will be less about the tools we have at our disposal and more about the skills and knowledge we amass through the network to help us wield them.

Anywhere Work

Over the centuries, technology has helped automate the largest parts of the workforce: first unskilled labor, then skilled labor in factories, then white-collar workers. Anywhere is the next set of chapters following the shifts from the telegraph, telephone, adding machine, computer, and the Internet. Connectivity is allowing us to further automate knowledge work, freeing up yet more potential productivity in the workforce over time. (See Figure 12.1.)

Thomas Friedman's book *The World Is Flat* set out in clear relief the changing roles of countries now able to participate in a global economy. The flattening process actually began as container shipping made offshore manufacturing economical, bringing inexpensive goods from Japan, Taiwan, and Singapore to eager customers in new Western markets. But as Friedman explains, undersea cabling then brought the cheaper knowledge workers of India, Philippines, and Indonesia online too, creating a broader, richer, and more complex global economy.

The emergence of the technologies we're discussing has coincided with the emergence of more service-based regional economies. Telecommunications in general and the Internet in particular con-

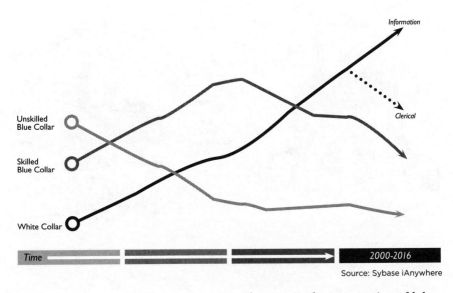

Source: Sybase iAnywhere

Figure 12.1 *Changes in the automation of labor.*

tributed heavily to markets like the United Kingdom becoming (for better or worse) very dependent on the service sector for growth. That trend may continue as connectivity becomes ubiquitous.

In the world of work, Anywhere will lead us to a postglobalization business era. Beyond moving entire business operations to less expensive labor markets, as we've seen with the rise of offshoring and outsourcing, the next phase will be what I call the "atomization" of work, the ability to shift business activities to other locations at a more elemental level and to do it dynamically—following the sun, or foreign exchange rates, or the vagaries of war or weather. The Anywhere workday will be filled with granular and dispersed activities, doled out and reassembled by leaders who will manage projects and results in very different ways from the way we do now. (See Figure 12.2.)

"If you think about business globalization today," confirms Mark Templeton at Citrix, "It's been about companies taking a really big chunk of something and moving it somewhere else. That's a first-generation thought, just like client-server computing was a first-generation thought about moving some computing from a central server to the client and networking them together." But in the Anywhere world, he agrees, "We'll be able to pass business process from location to loca-

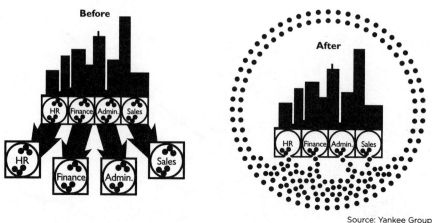

Source: Yankee Group

Figure 12.2 *How the distribution of work changes.*

tion in new ways. The old way was to move the entire order-to-cash group somewhere else. In the future, order entry will be in one place and collections in another, optimized at a lower level of granularity."

This atomization won't come easily. How will we collaborate to get work done in such small pieces, by many different people, in constantly shifting locales? Our command-and-control business structures, derived from medieval conventions in organized religion and the military, will need to evolve. "Mostly it's been in just the last 50 years that we have thought about business structure the way we do today. But it will change," says Sybase's Terry Stepien. He believes that the enterprise will transition from today's rigid structure to one more like a community. "In the film industry, a small core of people round up other experts to make a film. Then they go away and recombine to do the next one. An enterprise will be a collection of people who come together to work on things."

Sogeti CTO Michiel Boreel isn't so sure. "If you only care about each activity in the network as an independent element, that might be right. You can create markets, outsource it all, and become the entity that just organizes the work. But when there's more value in what connects the activities than in the activities themselves, you will choose to do them inside an organization," he says. "Organizations will change shape, I think, but you'll see both models, where things are done outside the company or still inside."

How much value does your firm create in between the atomic-level efforts that make up your activities? And what will doing something "inside" the enterprise mean? If your workforce is contracted on more flexible terms than captive full-time employees, is that "outside"? Part of your transition to an Anywhere Enterprise will involve finding that balance.

As organizational structures evolve, so must management tools and techniques so that they reflect the loss of direct control over employees and instead use new techniques for achieving results in a more fluid staffing context. LiveOps has built a performance-based meritocracy algorithm for the scheduling and routing of calls to independent agents, who take calls on behalf of companies. This means that top-performing independent agents get to pick their work schedule first, and make more money because more calls are routed to them. This has many similarities to the methodologies found in gaming as well as community-based businesses like eBay. The independent agents who contract to LiveOps sign up for time slots to receive incoming calls to handle based on their availability, performance, skill set and program certifications. "We don't tell them what to do; instead we set up conditions that guide them toward what we're hoping for. Then we adjust quickly if it doesn't work," explains CEO Maynard Webb. How will your firm move to managing outcomes instead of bodies?

For some of the executives I spoke with for this book, the imperative to organize and manage differently highlights how unprepared their tools are. Stepien points out that the consumer world is quite far ahead of the enterprise in using the network to collaborate. "If you think about it, all the social sites are really collaboration sites. That concept hasn't become mainstream in the working world yet. It's not zero—there are things happening. But today, what most enterprises use to collaborate in place of Facebook and better tools is nothing more than big, long e-mail threads with copy lists. We have a long way to go."

Here Boreel agrees. "People are changing faster than their organizations are. Sogeti is giving them collaboration tools as quickly as possible as a result. If we don't, they'll find some random public site called www.letsworktogether.com, and start sharing all kinds of confidential stuff. Who's hosting it? What are the privacy and security policies? We have to anticipate that." Will your enterprise do the same?

Innovation Comes from Anywhere

ENIAC was the world's first general-purpose computer, built for the U.S. government to calculate artillery firing tables. When I was an engineering student in the late 1970s at the University of Pennsylvania, where ENIAC had been constructed 30 years earlier, its dusty remnants lay nearly forgotten in a dimly lit basement. A few of us once paid it a furtive visit to marvel at its bulk and peer at the Bakelite dials and peeling labels. Thankfully it now sits in a museum, revered for the milestone it represents in applying digital computation to the complicated mathematics of a battlefield problem. In computing's earliest days, it was originally the needs of the military—massive and urgent—that spawned and funded technology innovations, which then moved to the commercial arena as the innovations stabilized and scaled.

The sources of technology innovation are also shifting with the Anywhere Revolution. Since the emergence of the personal computer, the balance of technology's early adoption has progressed steadily. Now it's the consumer market providing the impetus for new ideas. The expenses associated with launching new technologies into the mass market are still considerable, but the worldwide consumer appetite for technology innovation is increasingly voracious, so the potential volume of broad-based successful consumer products could mean a big payoff. The consumer now leads the way in adoption of portable devices, data and software in the cloud, and network-enabled collaboration; companies now follow.

But the Anywhere Revolution promises not only to preserve that recent shift in early technology adoption from companies to consumers, but to help innovation's pioneering forefront to move from mature markets to emerging markets.

Developed markets have been leaders in innovation around the world for so long that it's almost reflexive to assume it will always be this way. Mature markets have had the schools, the investment funds, the local talent, and the largest source of buyers for those innovations.

Those creations—computers, phones, more—only progressed to emerging markets as the economics improved, as costs dropped enough to make them viable for less well-funded regions, as import and export regulations allowed, and as those markets realized the potential they offered and generated enough demand to pull them in.

Friedman's 2005 book pointed to the millions of newly minted engineers joining the working world from the thousands of universities in Asia. But two recent shifts in the global landscape will further affect the expansion of Anywhere technologies: the economic crisis that began in 2008 and the awakening of the developed world to affordable mobile connectivity. These two factors will change the innovation game in a fascinating way.

The timing and nature of the recovery from the recession that blanketed the world at the end of the last decade may be no more predictable than its arrival. But many anticipate that mature markets will come out of recession slowly, held back by a hangover from the surplus consumption of companies and consumers in the prior decade. In technology adoption, firms just recovering from depressed demand are likely to make new innovation investments cautiously. Venture capital partnerships, looking at reduced investment funds and lower probabilities of successful exits from their investments due to fewer initial public offerings, will operate more conservatively, potentially constraining the number of new firms that get off the ground with investment funding rather than through bootstrapping their operations. It may take a generation for these conditions to ease.

Meanwhile, emerging markets exhibit rampant demand for communications technologies and all the other resources that can be brought to their shores with first-time connectivity. IBM Global Financing is the largest IT lender in the world; its general manager John Callies says, "Emerging markets don't want our leftovers. They want current, contemporary solutions." The home-grown talent and capital in these markets will be encouraged by governments that clearly see economic impact from the results.

Most important, they will have to tackle issues that are indigenous to their markets—and their solutions will become available elsewhere. "Power will soon be the number one gating factor in how large and wide you can build a network—no matter where it is. The power problems that the developing world has to solve now will come home to roost in the West, too," says Vanu Bose, the wireless technology entrepreneur. He points out the shift ahead as a result. "It's an interesting reversal. Emerging markets today face industrialization, pollution, all the things we are dealing with in mature markets, but *after* we faced them. Now they're hitting problems that we haven't gotten to yet." Dov Baharav at network software leader Amdocs agrees: "In many places in emerging markets, they are finding a way to implement communications systems in a much cheaper way. They're ready to pay, but they need to do things differently. Some of those practices and new ideas will find their way to mature markets later on."

Mike Muller at chip technology company ARM chimes in: "The classic Western white man stereotype is not where the action is going to be. Rural India may bypass the PC entirely. Once you have Google on your phone with a keyboard, why would you buy a PC? Part of the world will sidestep some of our technology evolution."

If you aren't already operating some portion of your business in an emerging market and if you aren't already getting to know the customers and problems there, will you spot the innovations emerging there in the next decade at the right time?

Anywhere Products: From Swords into Plowshares, Right Now

In pure technology terms, the world of things around us is also going through a fundamental change. I think Mike Muller puts it best: "It used to be that what you made was what customers actually used. Like a washing machine: five years after you sell it to them, they're still going to be doing the same thing with it. But we're moving to a world of products that are more like computers. As the company who sold it to you, we may have no idea what you're going to do with it— either now or in five years."

"Morphing" is the term that computer animators use to describe a transition from one image to another, right before our eyes. How might your products morph? How will people want them to develop and change in the future? How are their needs going to change and how can your devices meet those changing needs? Consumers have become conditioned to expect a steady wave of new features. Regardless of what that future morph will be, a connected product comes with the path to that morph built right in. Resculpting the contours of a product's function for the user will be a responsibility the seller shares with the buyer: the buyer letting the seller know what he wants from it, and the seller delivering the new function through that persistent connection back to the buyer. Don Tapscott emphasizes two very relevant traits of Next Gen consumers that should boost your resolve: they expect to be able to customize—to make a product their own—and to collaborate with a seller to help make the product or service better. I think about the change as going from selling someone a paint-by-the-numbers kit to giving her the brushes, the paint, and a blank canvas to create what she likes. (See Figure 12.3.)

To stay competitive, you'll need to think about your product or service in this more fungible sense: that of selling the customer a vessel through which you change various specific functions for new ones as needed. Success will depend on openness with customers and a willingness to learn along the way. Businesses will learn a lot from their consumers—not from stale devices such as postsale surveys and warranty cards, but instead through the ongoing use of connected

Before Anywhere   **After Anywhere**

Source: Yankee Group

Figure 12.3 *Customers take the reins.*

devices and experiences. Using connectivity experience data embedded in those devices' use of the network will be key to understanding and meeting your customers' appetites.

But you will have to be quick about it. New product expectations, trial, and acceptance have accelerated to operate almost at the speed of the very networks we use to make things happen. As the network creates instant awareness, Anywhere lets us leapfrog the old, "I tell two friends, and then they tell two friends" word-of-mouth phenomenon for product adoption.

The need for speed won't negate the requirement for a thoughtful approach. Mindlessly bolting on connectivity, like a toolshed on the side of a house, could fail. Connectivity is more than a gimmick, a new cup holder for next year's minivan. A better approach is one that rethinks your customers' product experience with connectivity at its center and redesigns the product to incorporate it in the most suitable way. How will your firm be quick, agile—and how will it plan?

The Anywhere Society

As ubiquitous connectivity brings us together, a totally new context for relationships and experiences comes with it, not just commercial relationships but social ones as well. Ten years ago, it wasn't common for people to date online; couples were reluctant to admit that they met on Match.com or Craigslist. Now, it's completely acceptable to have cyber relationships. We make new friends and stay in touch with everyone online instead of picking up the phone or writing a letter. And these online activities all feel normal.

Anywhere is redefining the social fabric of the twenty-first century. We're all struggling with what that means. As I write this, the government of Ishikawa prefecture in Japan is attempting to ban the use of cell phones by children under 15. But meanwhile, the Stargate theme park in Dubai is equipping children with a wireless sensor locket during their visits so that the park and their parents always know where they are.

Connectivity is redefining what it means to be alone and to feel alone. Can we be alone when we are constantly connected? Loneliness and isolation will be less about our physical location and more purely about our emotional state. Milo Medin, cofounder of wireless broadband start-up M2Z Networks, believes that may be the most important mandate that connectivity has. "My cofounder was an immigrant to the United States, as were my parents. English was his second language and mine too. We both know that sense of being disconnected, being outside of things. Being isolated from the Internet reminds me of how I felt in school—not knowing what was going on."

As a child, I had a neighbor who was an amateur ham radio operator. We'd sit in front of a bank of equipment in his basement, and I'd watch as he patiently twisted knobs and flipped switches. Eventually the muffled, crackling voice of his radio buddy in western Australia would emerge into the heavy earphones on my head while I stared at a faded photo of the man that was tacked up on the wall in front of me.

Today we can while away hours online in chat rooms, either pretending to be someone else, or blessedly free to be who we feel we really are. And for many of us, off-line is the new luxury, with messages from exotic resorts promising no phones, no Internet! Get away from it all! The Internet brings us news of the tragic actions of society's most extreme social outcasts gunning down strangers, but how often does it silently rein in other outcasts, letting them find a friend and saving them from that final terrible expression of isolation?

As the fabric of our society changes, its administrative elements are changing, too. Ubiquitous connectivity will eventually transform the way we organize ourselves and govern our societies. Grassroots organizations are popping up online to champion political and social causes from voters' rights and transparent election results to education and an end to poverty. Most of these efforts are relatively disjointed now, but what's probably ahead is a more strategic maturation: cohesion and collaboration. There could ultimately be less of a role for centralized government institutions with fixed physical infrastructure and more opportunity for individual participation via virtual groups organized in the cloud. Early signs of this were visible in the historic 2008 campaign for the U.S. presidency, which took grassroots organizing online and to the mobile platform to a level never before employed. Following the election, the architects of that initiative set out their next goal: to transform that online success into a new form of ongoing citizen input and outreach throughout the life of the new administration.

Anywhere as a Right

Zooming back from the Anywhere world even farther, it becomes clear that a twenty-first century expectation is emerging. Consider this: Connectivity can bring the world's riches—health, love, education, employment, material goods—to all of us. It can also bring a powerful knowledge of alternatives to what would otherwise appear to be our lot in life, telling us about others who feel the same way, empowering us to rise up over tyranny. Why wouldn't the network thus be as vital to us as food and water?

The playing field is leveling out as people all around the world get access to a global collection of resources and as the terms on which human beings transact with each other are being rewritten. A new human right is emerging, perhaps the most fundamental right of all: the right to be connected to the world. When connectivity is all we need to provide for ourselves and our family, how can we justify letting anyone go without it?

Dov Baharav at Amdocs calls wireless an implied human right. Ben Verwaayen at Alcatel-Lucent agrees: "Brains have been reliant on where they've been born to have a chance to monetize themselves. But if you want people to have equal opportunity, they need to be able to participate in the global economy, so it shouldn't matter where they are. If that's true, then there is only one requirement for that to happen: access to the network."

Anywhere and You

Zoom way, way in now: from the big, world-circling implications on economics; work and products; society, government, and human rights; all the way back down to your own personal implications. Let's talk about how you will respond to the challenges and opportunities that the emergence of global connectivity presents to you as a citizen and as a businessperson.

You can expect to encounter obstacles inside your own firm. Mark Templeton at Citrix warns us about the biggest internal risk. "There is a lot of inertia around what I call dead ideas," says Templeton. "So many things we deal with in business were built on assumptions and facts that were true 25 years ago." Inertia in the face of change is frequently identifiable in an organization by repeated self-talk that dismisses external threats. If you hear your colleagues convincing themselves why the advance of connectivity isn't going to change what your firm offers, or how it gets that accomplished, alarm bells need to go off.

Find some like-minded colleagues and consider staging an intervention to snap the group-think denial. Practice saying, "But what if we're wrong about that? What will we do then? Shouldn't we have a plan?" Prepare your counterattack by pointing to other firms in your industry that were overtaken by previous market events: continuing to sell buggy whips as consumers moved to the automobile, resisting the call, from buyers who'd moved online already, to present their wares to them electronically, or ignoring breakthrough business model efficiencies from disruptive low-cost competitors that ultimately sank their profits.

A related negative force you'll have to fight off is incrementalism. "No one ever starts with a fresh piece of paper," continues Templeton. "Incrementalism means people focus on fixing legacy things that are broken. What that does is give you layers of things that just propagate what the past has been about." We mention the early days of television earlier in this book and how transformational they were to other industries like advertising. But in the first days of programming, the thinking about what to broadcast on television was just as incremental as Templeton complains. Programs consisted of announcers from radio, now sitting in front of both mikes and video cameras, doing the same thing they'd done in their radio program. It was some time before exciting television programming emerged that capitalized on the intrinsic advantages of imagery over sound alone—and it emerged from the minds of people willing to look at the new medium with fresh eyes.

Today, early forays into mobile advertising exhibit the same one-dimensional thinking. Advertisers use formats from and make assumptions based on current advertising channels. Then they shoehorn them into the pocket-sized mobile experience. Why on earth do we think it will work? The theory is compelling: consumers want a subsidized mobile experience and will respond to brands that help provide it. But until we break free of some dead-idea baggage to find the uniquely appropriate ways to use the world's most personal technology to talk to consumers, it won't get off the ground.

If you manage to keep your firm from talking itself out of concern about the Anywhere Revolution and you come up with some fresh thinking about how to approach it, beware the temptation to try to do it all. Encourage coworkers to think in Anywhere stages, progressively expanding markets, expanding broadband consumption, increasing connectivity-based functionality, and more. Keep things simple, tee up lots of follow-on possibilities, and let your Anywhere adventures' first successes help shape the best next steps.

That's change advice for your company. As a businessperson, how will you let Anywhere change *you*? I have three final thoughts.

1. Be Anywhere Curious

The analysts at Yankee Group look everywhere for details of the expansion of connectivity and clues about what it will mean and how to put it to work for our clients. You should develop your own keen sense of the strengths and weaknesses of connectivity technologies and what would fit well with your customers' appetites and your enterprise's culture and personality. You can't predict a black swan—but you should be looking for them all the time.

That's why Bob Metcalfe, general partner at Polaris Venture Partners, started using Twitter early. "New things are an easy target for those who lack imagination. Having met a thousand sensible people who told me for a decade that my network technology was doomed and then being proven right, I'm now handicapped forever. When I am surrounded by sensible people, I reject everything they say," he asserts with a wry smile. "Years ago, no one understood why e-mail was worthwhile. Now, no one thinks twice about it—but they're busy talking about why Twitter is stupid."

You might not have the confidence borne of Metcalfe's personal successes, but you can still be an eager investigator, avoiding the temptation to write things off before you've opened your eyes to all the possibilities they offer. Don't let yourself be talked out of powerful ideas.

2. Be an Anywhere Connector

Malcolm Gladwell's book *The Tipping Point* introduced the idea that some people are naturally "connectors," using their links with others to create or promote opportunities between them. Anywhere dramatically reduces the cost of being a connector. Julie Woods-Moss, formerly president of strategy, marketing, and propositions at British Telecom, says: "If the barriers to connecting are high, few people will do it. Whereas if you know how people want to connect with you and it's very easy to choose the right method, why wouldn't you do it? If I can help someone with a one-minute effort with no downside to me, then I should. Networks are self-nurturing; the more you feed them, the stronger they get." Yankee Group analyst Declan Lonergan points out that ubiquitous connectivity might reduce the unique value of Gladwell's original connector-type people because we can all be connectors. If that's true, then Anywhere will force us all to raise our game.

3. Be an Anywhere Thinker

When I was struggling through high school chemistry, I was trying to memorize the periodic table of the elements when I challenged my chemist father to recite it for me. "I don't need to know it by heart," he said. "I know where to find it." Even more true today and in our Anywhere future, your skills in searching for information and then synthesizing that information, looking for patterns, and interpreting it will become much more valuable than the actual information you've amassed in your experience as a person and as an employee.

Metcalfe predicts that the biggest impact of ubiquitous connectivity will be the facilitation of *collective intelligence*. It's an exciting thought. Anywhere enlarges the potential sources of input for virtually all decisions; that should bring tremendous change to how you approach your own thinking and decision making. In medicine, doctors today act based on what they learned in medical school or what they have personally encountered in their practice. Eventually, decision support service providers in medicine will aggregate knowledge based on all known evidence. Any one doctor will be able to know what *all* doctors know.

In 2008, Nick Carr wrote a provocative article for *Atlantic* magazine asking "Is Google Making Us Stupid?" In it, he wondered whether the ease of getting answers was making us intellectually lazy. I think it's a thoughtful question, but the wrong one to be asking. You should be asking yourself these two questions instead: How can the universe of human knowledge, soon to be available anywhere instantaneously, make you *better*? To improve how effectively you work with mountains of data, what personal growth and development do you need to undertake?

Parting Thoughts

I was an adolescent in 1973 when my family and I were camping our way around Europe. On the Greek island of Corfu one night, I sat at a picnic table at the water's edge, our kerosene lantern flickering in the night breeze. Peering out at the darkened surf, we could see a few faint lights twinkling back at us from another shoreline close by. My mother explained that it was the country of Albania, isolated from the rest of the world for decades by its tyrannical leaders. My young romantic mind was fascinated that something just two tantalizing miles away was closed off—mysterious, unknowable. I peppered her with questions she couldn't answer: what language do they speak, what do they eat, what do they do? Frustrated but captivated, I sat and stared. Was there at that very moment maybe a young girl like me in Albania, looking out over the water back at us, also trying to imagine who we were, what we were like?

Today's Albania is still just two miles from Corfu, but it's a very different place. Now a member of NATO and a candidate to join the European Union, the country has opened up to the rest of the world. Ferries run tourists back and forth to the port town of Saranda from the beach near where I wondered about a possible doppelganger on the far side of the water.

The romantic mystery of a closed Albania may be no more, but to its inhabitants, I'm sure it felt less like poetry and more like prison. The opening of the world, the sharing of its riches with everyone, is the heart of the promise of our Anywhere future.

I met an executive recently who happened to mention that he left his old job for a new position at Google because it offered him a chance to change the world. The vision of Anywhere thrills me because it gives *all of us* that chance. When worries about technology's challenges want to rain on my Anywhere parade, my enthusiasm is renewed by the imaginative and persistent innovators around the world who will make Anywhere happen.

For me, Nicholas Negroponte captures the true potential of Anywhere with this story: "In Cambodia in 2001, I took a photo of the first kids we equipped with laptops, all holding them up proudly for the camera. Eight years later, every single one of those kids was still in school. I always say, I'm not in the PC business—I'm in the hope business."

Operate with enlightened self-interest. What can you do to make the difference between Anywhere good and bad? How can you ensure you reap the benefits and avoid the dangers? I've focused this book on explaining how the advances ahead in connectivity will revolutionize your business, and how that transformation can be profitable if you take the right steps. But the best thing you can do? Be an optimistic adopter of connectivity—promote the positives and push for change. Be an Anywhere evangelist and embrace the Anywhere future that is opening before you.

NOTES

For more information on the Anywhere Revolution, including deeper interviews with some of the experts consulted in the research for the book, please see the book's Web site, anywhere.yankeegroup.com.

Where sources are quoted more than once in a chapter, only the first instance is referenced below.

Chapter 1

Page 3. "The Internet gave us time freedom...." Hilmi Ozguc, founder, Maven Networks. Personal interview, April 24, 2009.

Page 7. "Vitality, Inc. has developed a pill bottle...." David Rose, founder and CEO, Vitality, Inc. Personal interview, March 25, 2009.

Page 11. "Starting from the launch of the Netscape browser...." K. G. Coffman and A. M. Odlyzko, "Growth of the Internet," *AT&T Labs*, July 6, 2001.

Page 14. "Historian Ralph Harrington explains that people...." Ralph Harrington, "The Neuroses of the Railway," *History Today*, July 1994.

Chapter 2

Page 25. "This revolution can't be stopped...." Ben Verwaayen, CEO, Alcatel-Lucent. Personal interview, April 7, 2009.

Page 31. "We're starting to see the Internet effectively become television...." Paul Sagan, CEO, Akamai. Personal interview, May 1, 2009.

Page 32. "But in Japan ..." Dr. Botaro Hirosaki, NEC. Personal interview, June 29, 2009.

Page 32. "The definition of broadband won't be static ..." Walter B. McCormick, CEO, United States Telecom Association. Personal interview, July 1, 2009.

Chapter 3

Page 41. "Communications will become global ..." Reed Hundt, former chairman, U.S. FCC. Personal interview, June 1, 2009.

Page 43. "... network ubiquity is inevitable. That's because the utility of any network is in its ubiquity ..." Walter B. McCormick, CEO, United States Telecom Association. Personal interview, July 1, 2009.

Page 44. "... similar uncertainties arose during the buildout of electrical infrastructures in the early 1900s." Nicholas Carr, *The Big Switch: Rewiring the World from Edison to Google*, W. W. Norton & Co., Inc. 2008.

Page 55. "The more prosperous you are, the more connectivity you can afford ..." Dr. Robert Metcalfe, general partner, Polaris Venture Partners. Personal interview, April 23, 2009.

Chapter 4

Page 63. "All devices will be connected; they'll have to, to perform the work we want them to." Glenn Lurie, president, AT&T Emerging Devices. Personal interview, April 20, 2009.

Page 65. "It wasn't too long ago that people were saying, 'eBooks? Whatever.'" Steve Haber, president, Sony Electronics digital reader division. Personal interview, April 21, 2009.

Page 66. "... have a quick throw-down against a friend in the car ..." James Brightman, gaming expert. "The Future of Handheld Gaming," *Game Daily*, January 2005.

Page 67. "People are already beginning ..." Hilmi Ozguc, founder, Maven Networks. Personal interview, April 24, 2009.

Page 69. "Yankee Group's own consumer surveys in the U.S. pointed ..." For further information, see the Yankee Group Report, "How to Spot the Next IPTV Growth Market" by Vince Vittore, August 2009.

Page 69. "Weight scales, blood pressure cuffs ..." David Rose, founder and CEO, Vitality, Inc. Personal interview, March 25, 2009.

Pages 69–70. "Lots of things will be tried; some will turn out to be useful." Vanu Bose, founder and CEO, Vanu, Inc. Personal interview, May 4, 2009.

Page 72. "My favorite negative example is the 'spork' ..." David Rose, founder and CEO, Vitality, Inc. Personal interview, March 25, 2009.

Page 73. "The lens through which I look at brands ..." Tom Sebok, President and CEO, Young & Rubicam. Personal interview, June 26, 2009.

Page 73. "In the hospitality industry, Zagat was ..." Randall Stross, "How Many Reviewers Should Be in the Kitchen?" *The New York Times,* Sept. 6, 2008.

Page 75. "I spent an hour and a half on the Amazon site." Larry Weber, CEO, W2 Group. Personal interview, April 21, 2009.

Page 76. "There will always be stuff you won't do with your device ..." Mike Muller, CTO, ARM, Inc. Personal interview, May 14, 2009.

Page 77. "Kodak came out ... with a digital picture frame." Steve Tomlin, founder and CEO, Chumby Industries. Personal interview, April 21, 2009.

Page 80. "You could see Sprint partnering with ..." Dan Hesse, CEO, Sprint/Nextel, Inc. Personal interview, April 9, 2009.

Page 82. "Everything retailers do, they do for margin." Liudvikas Andriulis, CMO, Effortel. Personal interview, June, 16, 2009.

Chapter 5

Page 85. "We all just want to be a part of the world." José María Álvarez-Pallete, CEO, Telefonica International. Personal interview, April 14, 2009.

Page 90. "Mobile couponing is a new opportunity ..." For more information, see the Yankee Group report, "Advanced Payment Models of the Anywhere Network Will Create Opportunities," by Jon Paisner, August 2009.

Page 92. "The ultrasound technology described in this scenario is being trialed...." For further information, see the Partners in Health Web site: http://www.pih.org/.

Page 98. "In 2008, Stockholm ran its first pilot test ..." Tomas Bennich, president, Sweden Mobile Association. Personal interview, April 7, 2009.

Page 100. "As the saying goes 'it's expensive to be poor.'" Rajeev Suri, Nokia Siemens Network. Personal interview, April 13, 2009.

Page 100. "Western Union serves a group of people who in the past ... transactional channel of the future." Matt Dill, senior VP, Western Union Digital Ventures. Personal interview, April 8, 2009.

Page 101. "Nortura needs to unload trucks arriving ... from other centers." Are Bergquist, CEO, Matiq. Personal interview, June 12, 2009.

Page 102. "Mobile couponing is cheaper...." Lawrence Griffith, CEO, Samplesaint. Personal interview, June 18, 2009.

Page 103. "But given the intensely personal relationship...." Tom Sebok, President & CEO, Young & Rubicam. Personal interview, June 26, 2009.

Page 104. "... there will be many more 'screens' in the home...." David Rose, founder and CEO, Vitality, Inc. Personal interview, March 25, 2009.

Page 104. "The next compelling venues ... in the house." Steven Tomlin, founder and CEO, Chumby Industries. Personal interview, April 21, 2009.

Chapter 6

Page 105. "We want what we want, when we want it. We want to be in control." Larry Weber, CEO, W2 Group. Personal interview, April 21, 2009.

Page 114. "Online banking and e-commerce have been core...." Doug Brown, senior VP, mobile product development, Bank of America. Personal interview, May 12, 2009

Page 117. "Sony was late to enter the digital audio player market...." Saeed Shah: "How Sony lost the plot," *The Independent*, http://www.independent.co.uk/news/business/analysis-and-features/business-analysis-how-sony-lost-the-plot-487525.html, Jan. 21, 2005.

Page 117. "In 2004, *New York Times* reporter Ken Belson pointed out...." Ken Belson, "Infighting Left Sony behind Apple in Digital Music. Can It Come Back? Teaching an Old Walkman Some New Steps." *New York Times*, April 19, 2004.

Page 118. "Companies such as INQ have followed Apple's lead...." Steve O'Hear. "Hands-on Review: INQ1," last 100, March 23, 2009.

Chapter 7

Page 125. "Work is the new 'killer app' for computing." Maynard Webb, CEO, LiveOps, Inc. This quote and all others from Maynard Webb, personal interview, May 11, 2009.

Chapter 8

Page 143 "With connectivity as your platform, you can have a boundaryless enterprise." Doug Hauger, general manager, business strategy, Windows Azure, Microsoft. Personal interview, April 15 and April 16, 2009.

Page 149. "Drug maker Eli Lilly had an in-house computing environment...." Werner Vogels, CTO, Amazon. Presentation at the Telefonica Leadership conference, April 15–16, 2009.

Page 153. "If you look back 15 years ago...." Maynard Webb, CEO, LiveOps, Inc. Personal interview, May 11, 2009.

Page 154. "Big Bang turned us from the nicest club in the world...." Michael Marks, formerly of Smith Brothers, in the *London Telegraph*, "It started with a bang", Oct. 23, 2006.

Page 156. "Lots of equipment moves around; some things get lost...." Terry Wagner, CIO, SUNY Upstate Medical University. Personal interview, June 16, 2009

Page 157. "Anywhere you have parts, tools, or people as part of your workflow...." Antti Korhonen, CEO, Ekahau. Personal interview, July 14, 2009.

Pages 157–158. "At least a dozen years ago, Avis did something that all of us can connect with...." Russ McGuire, VP, corporate strategy, Sprint/Nextel. Personal interview, May 4, 2009.

Page 158. "In 2006, we got a now-famous letter from Walmart...." Christian Verstraete, chief technologist for manufacturing and distribution industries, Hewlett-Packard Co. (sidebar) Personal interview, June 12, 2009.

Page 163. "... only about 30 percent of our time, energy and dollars in IT were spent on differentiated value creation." Werner Vogels, CTO, Amazon. Presentation at the Telefonica Leadership conference, April 15–16, 2009.

Page 163. Ubiquitous connectivity makes work a verb...." Larry Weber, CEO, W2 Group. Personal interview, April 21, 2009.

Chapter 9

Page 167. "Mobilization of the world will continue exponentially." Rob Conway, CEO, GSM Association. Personal interview, May 14, 2009.

Page 171. "They all assume that a child will be plugged in with an AC adapter. It's not real." Dr. Nicholas Negroponte, founder and chairman, One Laptop per Child. Personal interview, April 2 2009.

Page 174. "All service businesses—banks, travel agencies, governments...." Axel Haentjens, senior VP of global strategy, France Telecom/Orange Business. Personal interview, May 20, 2009.

Page 177. "Leaders should just use the new stuff. It shows their employees...." Zeus Kerravala, senior VP, Yankee Group. Personal conversation, June 29, 2009.

Page 178. "Something like 60 percent of RIM's devices are now sold to consumers rather than businesses...." Josh Holbrook, director, Yankee Group. Personal correspondence, July 1, 2009.

Page 178. "Latency is evil ... adds up to a big number." Paul Sagan, president and CEO, Akamai. Personal interview, May 1, 2009.

Chapter 10

Page 183. "It would be a mistake to think there will be companies that use ubiquitous connectivity and ones that don't." Sriram Visnawathan, VP, Intel Capital and general manager, WiMAX Program, Intel. Personal interview, May 8, 2009.

Page 186. "Mobile devices are the only platform...." Nihal Mehta, founder, Buzz'd. Personal interview, June 29, 2009.

Page 188. "While 70% of India's population is rural and 80% of that group is illiterate...." Vanu Bose, founder and CEO, Vanu, Inc. Personal interview, May 4, 2009.

Page 189. "It's been estimated that nurses spend about 20%...." Dr. John Halamka, CIO, CareGroup. Personal interview, April 28, 2009.

Page 189. "We call it *unified asset visibility*...." Gabi Daniely, VP of marketing and product strategy, AeroScout. Personal interview, May 8, 2009.

Page 190. "We had a client with a fire at an offshore drilling platform." Michael Saucier, CEO, Transpara. Personal interview May 21, 2009.

Page 190. "The company's core service is for its traders." Rich Reiter, Bloomberg News. Personal interview, June 26, 2009.

Page 191. "What company doesn't want…." Maynard Webb, CEO, LiveOps, Inc. Personal interview, May 11, 2009.

Page 191. "No venture capitalist would have…." Werner Vogels, CTO, Amazon. Presentation at the Telefonica Leadership conference, April 15–16, 2009.

Page 192. "That's only part of the story…." Marc Benioff, CEO, Salesforce.com. Personal correspondence, May 6, 2009.

Page 192. "The cost per seat for installing, managing…." Tim Matthew, technology services manager, BAA. From a case-study of BAA's use of virtualization technologies published by Citrix, Inc. (see http://www.citrix.com/English/aboutCitrix/caseStudies/caseStudy.asp?storyID = 23693).

Page 193. "If there's one uber-concept in all this…." Mark Templeton. CEO, Citrix, Inc. Personal interview, May 25, 2009.

Pages 200–201. "How to Go Anywhere sidebar." All sources throughout sidebar as previously referenced, excepting Ron Sege, President & COO, 3Com Corporation. Personal interview, May 13, 2009.

Page 203. "Jim Collins *Built to Last* and many other books on corporate excellence,…" Jim Collins and Jerry I. Porras. *Built to Last.* (HarperBusiness, 1994).

Chapter 11

Page 207. "There will be a surprising and chaotic future to this." Dr. Robert Metcalfe, general partner, Polaris Venture Partners. Personal interview, April 23, 2009.

Page 209. "We can anticipate pretty well what technology will be available in the future...." Mike Muller, CTO, ARM, Inc. Personal interview, May 14, 2009.

Page 209. "As an industry, we're not a reliable source for knowing what people's behavior will be. We've gotten it wrong many times." Ben Verwaayen, CEO, Alcatel-Lucent. Personal interview, April 7, 2009.

Page 209. "The concept has gained in notoriety since Nicolas Taleb's book *The Black Swan* drew attention...." Nassim Nicholas Taleb. *The Black Swan: The Impact of the Highly Improbable.* (Random House, 2007).

Page 209. "Uncertainty doesn't mean you give up, that you can stop planning, scheming, and experimenting." Dr. Robert Metcalfe, general partner, Polaris Venture Partners. Personal interview, April 23, 2009.

Page 210. "Operators have to make money serving customers who only spend...." Rajeev Suri, Nokia Siemens Network. Personal interview, April 13, 2009.

Page 212. "With pay-per-megabyte pricing...." Declan Lonergan, VP, Yankee Group. Personal correspondence, July 12, 2009.

Page 212. "A new phone with a warranty in those markets...." Nigel Waller, CEO, Movirtu. Personal interview, July 10, 2009.

Page 214. "Building around an advertising-based business...." Chris Rothey, Navteq. Personal correspondence, July 12, 2009.

Page 215. "There will be new creativity in devices that we don't see today." Dan Hesse, CEO,

. Personal interview, April 9, 2009.

Page 219. "In India, the spectrum slices they're auctioning off are inefficiently small." Reed Hundt, former chairman, U.S. FCC. Personal interview, June 1, 2009.

Page 220. ...Yankee Group analyst Phil Marshall estimates the incremental cost on a per-user basis to be anywhere from $50 to $150 in capital expense.... For more information on the costs of moving to advanced 3G and 4G networks, see the Yankee Group Report, "Demystifying Long-Term Evolution on the Path to 4G," by Phil Marshall, July 2009.

Page 222. Yankee Group analyst Brian Partridge estimated in 2009 that telecommunications-related capital investment.... For more information, see the Yankee Group Report, "Global Telecommunications Capex Forecast," by Brian Partridge and David Vorhaus, March 2009.

Page 225. "Do you think it's politically feasible to create a national medical database?..." Dr. John Halamka, CIO, CareGroup. Personal interview, April 28, 2009.

Page 226. "Governments, well meaning as they usually are, tend to support the status quo." Dr. Robert Metcalfe, general partner, Polaris Venture Partners. Personal interview, April 23, 2009.

Page 226. "We need to find ways to encourage the fastest possible growth of broadband, particularly in the U.S." Paul Sagan, president and CEO, Akamai. Personal interview, May 1, 2009.

Page 228. "The relevant regulation today is oriented to banks and physical cash retail activities." Matt Dill, senior VP, Western Union Digital Ventures. Personal interview, April 8, 2009.

Page 230 (Sidebar). Personal interview with John Muleta and Milo Medin, cofounders of M2Z Networks, April 1, 2009.

Page 232. "The speed of innovation is not dependent on technology...." Michiel Boreel, CTO, Sogeti. Personal interview, June 17, 2009.

Chapter 12

Page 233. "The real meaning of ubiquitous connectivity is what it will mean to society as a whole to not have an isolated class."

Milo Medin, cofounder, M2Z Networks, Inc. Personal interview, April 1, 2009.

Page 234. A short but memorable film made in 1977 by Charles and Ray Eames called *Powers of Ten.…* For more information on the film, visit www.powersoften.com.

Page 235. "Connectivity is changing our world from a seller's market to a buyer's market.…" Paul Sagan, president and CEO, Akamai. Personal interview, May 1, 2009.

Page 236. "Healthcare was doctor-centric. With ubiquitous connectivity, it will become patient-centric.…" Dr. John Halamka, CIO, CareGroup. Personal interview, April 28, 2009.

Page 236. Thomas Friedman's book *The World is Flat* **set out in clear relief the changing roles of countries.…** Thomas Friedman. *The World is Flat: A Brief History of the Twenty-First Century*, (Farrar, Straus & Giroux, 2006).

Page 237. "If you think about business globalization today.…" Mark Templeton, CEO, Citrix Inc. Personal interview, May 25, 2009.

Page 238. "Mostly it's been the last 50 years." Terry Stepien, president, Sybase iAnywhere. Personal interview, June 5, 2009.

Page 238. "If you only care about each activity in the network.…" Michiel Boreel, CTO, Sogeti. Personal interview, June 17, 2009.

Page 239. "We don't tell them what to do.…" Maynard Webb, CEO, LiveOps, Inc. Personal interview, May 11, 2009.

Page 241. "Emerging markets don't want our leftovers. They want current, contemporary solutions." John Callies, general manager, IBM Global Finance. Personal interview, July 9, 2009.

Pages 241–242. "Power will soon be the number one gating factor in how large and wide.…" Vanu Bose, founder and CEO, Vanu, Inc. Personal interview, May 4, 2009.

Page 242. "... In many places in emerging markets, they are finding a way to implement...." Dov Baharav, president and CEO, Amdocs, Inc. Personal interview, June 12, 2009.

Page 242. "The classic Western white man stereotype is not where the action is going to be." Mike Muller, CTO, ARM, Inc. Personal interview, May 14, 2009.

Page 245. "My cofounder was an immigrant to the United States, as were my parents." Milo Medin, cofounder, M2Z Networks, Inc. Personal interview, April 1, 2009.

Page 246. "Brains have been reliant on where they've been born...." Ben Verwaayen, CEO, Alcatel-Lucent. Personal interview, April 7, 2009.

Page 246. "There is a lot of inertia around what I call dead ideas." Mark Templeton, CEO, Citrix, Inc. Personal interview, May 25, 2009.

Page 248. "New things are an easy target for those who lack imagination." Dr. Robert Metcalfe, general partner, Polaris Venture Partners. Personal interview, April 23, 2009.

Page 248. "... some people are naturally 'connectors'—using their links with others to create or promote...." Malcolm Gladwell's book *The Tipping Point* (Back Bay Books, 2002).

Page 249. "If the barriers to connecting are high, few people will do it." Julie Woods-Moss, formerly president, strategy, marketing and propositions, British Telecom. Personal interview, May 14, 2009.

Page 249. "... ubiquitous connectivity might reduce the value of Gladwell's original connector-type people." Declan Lonergan, VP, Yankee Group. Personal correspondence, July 23, 2009.

Page 249. "In 2008, Nick Carr wrote a provocative magazine article." Nick Carr: "Is Google Making Us Stupid?" *The Atlantic* (July/August 2008).

Page 251. "In Cambodia in 2001, I took a photo of the first kids...."
Nicholas Negroponte, founder and chairman, One Laptop Per Child.
Personal interview, April 2, 2009.

Anywhere Terms

Anywhere: The emergence of ubiquitous connectivity and its effect on consumers, businesses, and the world.

Anywhere Brand: A company whose brand value is recognized by consumers across many channels—including print, TV, online, and mobile.

Anywhere Consumer: An individual unfettered by the shackles of time and place who connects to content, society, and commerce at any time from anywhere.

Anywhere Economy: A geographical area with at least one broadband line per person.

Anywhere Enterprise: An organization whose employees, customers, assets, and partners connect to information and services when and where they need them.

Anywhere Experience: A networked experience that blends the quality of wired access with the ubiquity of wireless access.

Anywhere Index: The number of wired and wireless broadband lines per person in a country or region, expressed as a percentage.

Anywhere IT: Information technology that enables any user to work from any location, over any device, with the best possible experience that the device and network allow.

Anywhere Media: Media delivered to consumers when and where the consumer will pay attention to it.

Anywhere Network: A seamless intelligent network with broadband capacity and wireless ubiquity.

Anywhere Network Economy: The annual revenue generated by business and consumer broadband connectivity and the network hardware and software purchases each year that deliver those broadband connections.

Anywhere Revolution: The massive transformation the world is undergoing to achieve the Anywhere Network.

Anywhere Tipping Point: The point in a region's growth in connectivity when its broadband lines equal its population.

Anywhere Worker: An employee who is able to interact, collaborate, and conduct business with other employees, customers, and partners anywhere, at any time.

Emerging Anywhere: Regions of the world with less than one broadband line per three people.

Transforming Anywhere: Regions of the world with at least one broadband connection per three people.

ubiquitous connectivity: Vision of a future state when a seamless global network connects all people and all the things we care about.

Technology Terms

1G, 2G, 2.5G, 3G, 4G: The generations of the family of wireless telephone and mobile telecommunications standards; the generations range from 1G, an analog telecommunications standard of the 1980s, to the emergent 4G, digital telecommunications standards that are currently being developed to expand the capacity and flexibility of mobile broadband networks.

access device: An item that incorporates connectivity technology to use a network to send and receive data: a PC, a phone, a consumer electronics product, and, soon, many other devices in our lives.

access technology: The method used by a network user to transfer data (voice, media, etc.) to and from a device—either wired or wireless.

average revenue per user (ARPU): A common measure in the communications industry for how much money is made from a customer with a mobile phone. The higher the ARPU, generally, the more profitable the customer is likely to be.

bandwidth: The capacity of a network line to move data, expressed in thousands (K), millions (M), or trillions (G) of bits per second (bps).

Bluetooth: A wireless technology that is a short-range communications system intended to replace the cables connecting portable and/or fixed electronic devices.

broadband: Generally, a network connection with enough capacity to transmit multimedia data. Specifically, a wired connection that has a capacity of at least 384 kbps downstream, or a wireless connection using 3G technology or better.

bundling: The packaging of two or more telecom services together, for instance fixed-line voice and wireless voice, or fixed-line voice and broadband.

cloud computing: A style of computing in which dynamically scalable and often virtualized resources are provided as a service over the Internet. Users need not have knowledge of, expertise in, or control over the technology infrastructure in the "cloud" that supports them; both the software they use and whatever data that software manipulates reside somewhere in the network, rather than on the users' own computers and storage devices.

digital subscriber line (DSL): A technology that transmits digital data over telephone wires at a higher frequency than normal phone lines, enabling consumers to have Internet access over their phone lines even as they are still used for voice calls.

downstream: Describing data traveling on a network to its recipient; for instance, a user with a PC or other access device.

fixed lines: Phone lines usually run through copper or fiber for fixed locations such as the home or office; also referred to as land lines or main lines.

fixed/mobile convergence (FMC): The trend toward seamless connectivity between fixed and wireless telecommunications networks. The term also describes any physical network that allows cellular telephone sets to function smoothly with the fixed network infrastructure.

gigabits per second (gbps): A unit of measure of a network's capacity to move data. A gigabit is a trillion digital bits, or 1,000 times greater than a megabit.

hot spot: A wireless LAN (local area network) node that provides Internet connection and virtual private network (VPN) access from a given location.

information and communication technologies (ICT): An umbrella term that covers all advanced technologies in manipulating and communicating information.

IP, TCP/IP: Internet protocol; IP is part of the larger Internet protocol suite (shortened to TCP/IP), which is a set of standards for packaging data for transmission over the Internet and other similar networks.

kilobits per second (kbps): A unit of measure of a network's capacity to move data. A kilobit is 1,000 bits.

latency: The delay, usually expressed in milliseconds, in data arriving at its destination. Excessive latency creates difficulties for many connectivity experiences, including media consumption, financial transactions, and more.

location-based services (LBS): A service that can identify the user's location and incorporate it into information or transactions, such as mapping, banking, or commerce.

long-term evolution (LTE): A mobile broadband standard that is a migration path to 4G.

megabits per second (mbps): A unit of measure of a network's capacity to move data. A megabit is a million digital bits.

mobile Internet device (MID): A multimedia-capable handheld computer with wireless Internet access, generally larger than a smartphone but smaller than a tablet PC.

narrowband: A network connection with limited capacity, suitable only for text transmission and simple interactions.

near-field communications (NFC): Short-range wireless communications technology; in some retail applications it allows consumers to wave a card or device with an NFC transmitter in front of a receiver to accomplish payment transactions, for instance.

net neutrality: A principle that Internet traffic should be treated equally.

network intelligence: Network infrastructure that provides greater visibility into network, subscriber, and application activity that is provided by a regular telecommunications infrastructure.

open networks: Networks that allow all users to have unbundled and equal access to predetermined basic network functions and interfaces; as a result, users can personalize their experience and are not tethered to certain service providers or applications already on devices; also, networks in which internal information about network traffic, network users, and more is made available to services that use the network that are not operated by the network provider itself,

optical fiber: A glass or plastic fiber that carries light along its length. Fiber-optic communications permit transmission over longer distances and at higher bandwidths (data rates) than most other forms of digital communications.

over-the-top: Services, typically home entertainment, that are delivered across the broadband network but are not associated with a particular broadband service provider; for instance, watching YouTube is an over-the-top video service.

prepaid, postpaid: The two types of mobile phone plans; customers either pay for their cell-phone minutes prior to use (prepaid) or pay for their cell-phone minutes after use (postpaid).

radio-frequency identification (RFID): A small chip (typically referred to as an RFID tag) applied to or incorporated into a product, animal, or person for the purpose of identification and tracking using radio waves.

seamlessness: A quality of networks or technologies that has consistent interfaces and architectures and provide the user no disruptions or fluctuations in service quality.

simple message service (SMS): The technical term for the first and simplest transfer of data on wireless telephone networks, also known as texting or messaging.

smartphone: A mobile phone that offers advanced features like Internet access, e-mail capabilities, and assorted applications; smartphones have operating systems that are more integrated with the phone's features and main user interface than so-called "feature phones," which are characterized by proprietary operating systems and more limited functionality.

telecom: An abbreviation of telecommunications, the assisted transmission over a distance for the purpose of communication.

telematics: The receiving and sending of information via telecommunications devices; in recent years, this term has been most associated with global positioning systems (GPS), which integrate a computer with mobile identification technology.

triple play, quad play: A package of bundled services sold at a discounted rate by communications providers; among landline phone, mobile phone, broadband Internet access, and cable TV, a triple-play bundle includes three of the services and a quad-play offer bundles all four.

upstream: Describing data traveling on a network from its transmitter, such as a media company or a consumer sending messages.

user interface (UI): The graphical interaction with an application, device, or service experienced by a user.

virtualization: A broad term that refers to the abstraction of computer resources.

virtual private network (VPN): A computer network in which some of the links between nodes are carried by open connections or virtual circuits in some larger networks (such as the Internet), as opposed to running across a single private network.

voice over Internet protocol (VoIP): A general term for delivery of voice communications over IP networks such as the Internet or other packet-switched networks.

Web 2.0: The second generation of Web development and Web design, facilitating communication, information sharing, interoperability, user-centered design, and collaboration on the World Wide Web.

Wi-Fi: The name of a popular wireless networking technology that uses radio waves to provide wireless high-speed Internet and network connections.

WiMAX: A telecommunications standard for mobile broadband, moving mobile operators to 4G networks.

wired, wireless: Two ways to transmit information; wired means that wire lines are used to transfer information while wireless means information is transmitted using energy such as radio frequency, rather than wires.

Zigbee: A wireless standard protocol for radio frequency devices that uses the IEEE 802.15.4 radio specification and is particularly useful for low-cost, low-power wireless networks.

INDEX

Emily Nagle Green is the President and Chief Executive Officer of Yankee Group Research, Inc., a storied technology research firm that for over 40 years has guided a global clientele through profound connectivity change.

In her long and varied career in technology as an engineer, marketer, and researcher, and through her published work and speaking engagements, Emily has become widely respected as a thought leader on the impact of connectivity on networks, enterprises and consumers. She currently serves as the vice-chair of the Massachusetts Innovation & Technology Exchange (MITX), the U.S.'s largest Internet advocacy council. She was granted a B.S.L. cum laude in linguistics from Georgetown University and an M.S.E. in artificial intelligence and computer graphics from the University of Pennsylvania's Moore School of Engineering. Emily lives in Boston, Massachusetts, with her husband, Jack, and their daughter, Lydia.